Synthesis Lectures on Human-Centered Informatics

Series Editor

John M. Carroll, College cf Information Sciences and Technology, Penn State University, University Park, PA, USA

This series publishes short books on Human-Centered Informatics (HCI), at the intersection of the cultural, the social, the cognitive, and the aesthetic with computing and information technology. Lectures encompass a huge range of issues, theories, technologies, designs, tools, environments, and human experiences in knowledge, work, recreation, and leisure activity, teaching and learning, etc. The series publishes state-of-the-art syntheses, case studies, and tutorials in key areas. It shares the focus of leading international conferences in HCI.

Jeff Johnson · Austin Henderson

Conceptual Models

Core to the Design of Interactive Applications

Second Edition

 Springer

Jeff Johnson
Department of Computer Science
University of San Francisco
San Francisco, CA, USA

Austin Henderson
Rivendel Consulting and Design
Berkeley, CA, USA

ISSN 1946-7680   ISSN 1946-7699 (electronic)
Synthesis Lectures on Human-Centered Informatics
ISBN 978-3-031-50851-6   ISBN 978-3-031-50852-3 (eBook)
https://doi.org/10.1007/978-3-031-50852-3

This Springer imprint is published by the registered company Springer Nature Switzerland AG
The registered company address is: Gewerbestrasse 11, 6330 Cham, Switzerland

Paper in this product is recyclable.

Preface

We have been interested in Conceptual Models for many years. We both lived through the rough and tumble days of inventing the future at Xerox, and understand just how hard it has been (very) for the world to develop interactive applications that work even as well as they do. A continuing subject of discussion for many designers of interactive applications has been the "conceptual model of the system"—the view of the application that the designers hope people will adopt when using it.

Yet, there was a lack of clarity about exactly what conceptual models are. This is not entirely surprising, since so many different kinds of models and modeling are employed in the design and development of interactive applications.

For that reason, in 2002 we wrote an article for *Interactions* magazine—a 2500 word attempt to encourage designers to "begin by designing what to design". After it was published, we received feedback that the article was helpful to some designers and developers. Nonetheless, the use of conceptual models in software design and development remained fairly uncommon.

In the wake of the *Interactions* article, our own interest in conceptual models continued to evolve. We realized that there was more to say than we said in the article. We felt that conceptual models should be discussed more thoroughly and in a place that was readily available to all engaged in developing interactive applications or studying how to develop them. That was our motivation for the first edition of this book, which was published in 2011.

In the twelve years since then, the book has been used by many designers of user interfaces (UI), user experiences (UX), and even software. It has also served as the basis for many tutorials and courses on conceptual modeling that we presented to software design professionals. Several college professors, including one of us, also used the first edition as a textbook to teach conceptual modeling to college students—both undergraduates and graduates. Using the book in professional tutorials and college courses provided insight into where the book worked well, and where it did not.

Many who have used the first edition have provided us with feedback, either indirectly via ratings and comments at online booksellers, or directly via email and face-to-face at conferences. Some of the feedback was positive (see the first edition's ratings and comments at Amazon.com), and some of it pointed out deficiencies and ways in which the book could be improved. After Springer-Nature invited us to create a second edition, we compiled a list of improvements we wanted to make and solicited suggestions for improvement from others who had used the first edition.

In contrast to the first edition, the second edition:

- Summarizes early in the book the benefits of using conceptual models and how they fit into the design/development process.
- Expands the discussion of how conceptual modeling is related to other software design methods, and the explanations of what conceptual models are and are *not*.
- Defines a vocabulary for describing and discussing conceptual models, and adheres to that vocabulary throughout the book.
- Expands the explanation of "concept", "object", "attribute", and "operation" to try to overcome difficulties some designers—and design students—have in understanding those ideas.
- Updates outdated examples.
- Adds a second fully-worked-out realistic example of a conceptual model.
- Describes the user research methods that provide input for conceptual modeling.
- Clarifies the important components of a conceptual model.
- Adds a chapter on how to present conceptual models, specifically the results of an Objects/Operations analysis.
- Reorganizes and extends the discussion of conceptual modeling practices, discussing those practices that we consider essential and some that we consider optional and/or advanced.
- Expands the discussion of how conceptual models support collaboration between members of a design/development team.
- Briefly discusses what comes after conceptual modeling: proceeding from a conceptual model to the design of user interfaces.
- Adds an Appendix outlining the history of conceptual modeling and identifying sub-fields that have contributed to the practice.
- Provides chapter titles that are more self-explanatory than the first edition's cryptic one-word titles.

There are many people we would like to thank for their help in producing the two editions of this book.

We thank Jack Carroll for offering to publish the first edition in his *Synthesis Lectures on Human-Centered Informatics* series at Morgan & Claypool. His encouragement was supportive. We thank Diane Cerra for her help in setting direction and managing the mechanics of the first edition. We thank our colleagues Robin Jeffries, Jon Meads, Susan Fowler, Stuart Card, and Nigel Bevan for their helpful feedback on the first edition's first draft.

We thank Christine Killerich at Springer-Nature for inviting us to produce a second edition. We thank Hugh Dubberly, Prof. Ann Blandford, Jorge Arango, and Carola F. Thompson, for responding to our request for suggestions on how to improve the book, and Steven Pemberton for providing input on the Appendix. We also thank those who helped edit and design this book.

Finally, we each have people we would like to acknowledge:

Austin Henderson: I want to acknowledge many people for conversations over the years that have touched on conceptual models, particularly, Tom Moran, Stuart Card, John Rheinfrank, Shelley Evenson, Don Norman, David Asano, Jed Harris, and, of course, most centrally, Jeff Johnson. I want to thank my wife Lynne for her invaluable support and encouragement while this work has been underway, for two editions now.

Jeff Johnson: I acknowledge the many insights I have gained as a result of discussions—and arguments—about interaction design and conceptual models over the years with many colleagues with whom I have worked: Robin Jeffries, Bonnie Nardi, Steve Whittaker, Terry Roberts, Chuck Clanton, and of course, Austin Henderson. I also acknowledge the support and patience of my wife Karen Ande as both editions of this book were being written.

San Francisco, USA   Jeff Johnson
Berkeley, USA   Austin Henderson

Contents

About the Authors

Jeff Johnson is an Adjunct Professor of Computer Science at the University of San Francisco. He is also President and Principal Consultant at UI Wizards, Inc., a product usability consulting firm. After earning B.A. and Ph.D. degrees from Yale and Stanford Universities, he worked as a UI designer and implementer, engineer manager, usability tester, and researcher at Cromemco, Xerox, US West, Hewlett-Packard Labs, and Sun Microsystems. In the late 1980s and early 1990s, he was Chair of Computer Professionals for Social Responsibility. In 1990, he co-chaired the first Participatory Design conference, PDC'90. He has taught at Stanford University and Mills College, and in 2006 and 2013 taught as an Erskine Fellow at the University of Canterbury in New Zealand. Since 2004 he has served on the SIGCHI Public Policy Committee. He has spoken twice in the prestigious Authors@Google talk series, in 2013 and 2017. He is a member of the ACM SIGCHI Academy, a recipient of SIGCHI's Lifetime Achievement in Practice Award, and an ACM Distinguished Member. He has authored or co-authored many articles and chapters on Human-Computer Interaction, as well as the books *GUI Bloopers, Web Bloopers, GUI Bloopers 2.0, Designing with the Mind in Mind, Conceptual Models: Core to Good Design* (with Austin Henderson), *Designing with the Mind in Mind, 2nd edition, Designing User Interfaces for an Aging Population* (with Kate Finn), and *Designing with the Mind in Mind, 3rd edition, and Conceptual Models: Core to Good Design, 2nd edition* (with Austin Henderson).

Austin Henderson 50-year career in Human-Computer Interaction includes user interface research and architecture at
MIT's Lincoln Laboratory, Bolt Beranek and Newman, Xerox
Research (both PARC and EuroPARC), Apple Computer, and
Pitney Bowes, as well as strategic industrial design with Fitch,
and his own Rivendel Consulting & Design. Austin has built
both commercial and research applications in many domains
including manufacturing, programming languages, air traf-
fic control, electronic mail (Hermes), user interface design
tools (Trillium), workspace management (Rooms, Buttons),
distributed collaboration (MediaSpace), and user-evolvable
systems (tailorable—"design continued in use", pliant—"designing for the unanticipated" and ontologically extensible—
"scalable conversations"). These applications, and their development with users, have grounded his analytical work, which
has included the nature of computation-based socio-technical
systems, the interaction of people with the technology in those
systems, and the practices and tools of their development. The
primary goals of his work have been to better meet user needs,
both by improving system development to better anticipate
those needs, and by broadening system capability to enable
users themselves to better respond to new unanticipated needs
when they arise in a rich and changing world.

Introduction

Conceptual modeling as a tool/method to aid in designing software has a long history. Since the 1950s, computing pioneers and researchers have sought ways to specify software abstractly, in terms of the concepts it embodies and the tasks it supports, rather than the computing steps (e.g., code) required to implement it. In the decades since then, variants of conceptual modeling have been invented independently by researchers and practitioners in many subfields of computing, including database theory and operation, business-process specification, systems analysis, simulation, software engineering, very-high-level programming languages, information architecture, content management, semantic Web, task analysis and modeling, and user interface design. Despite differences in motivation, approach, and goals, many people in these subfields independently concluded that enumerating and modeling the important *concepts* of a software system before (or instead of) coding it in a conventional programming language facilitates development and improves ease-of-learning and ease-of-use. See the Appendix for a detailed history of conceptual modeling.

This book argues that a good Conceptual Model (CM) should be at the core of the interaction design of every digital artifact that people use to help them get their work done. This includes software products, electronic appliances, and digital services, but also products more generally.[1]

Those unfamiliar with interaction design often consider it to be designing the user interface or "skin" of an application: the controls, displays, error messages, use of colors, layout, etc. However, the interaction design goes deeper than the "skin": it includes all the concepts that the application's user interface exposes to users, and it includes the plans of activities that people perform to accomplish the tasks that they are using the application to

---

[1] For simplicity, this book uses the term "application" to cover all such artifacts.

© The Author(s), under exclusive license to Springer Nature Switzerland AG 2024
J. Johnson and A. Henderson, *Conceptual Models*, Synthesis Lectures
on Human-Centered Informatics, https://doi.org/10.1007/978-3-031-50852-3_1

support. The interaction design has roots that extend deep into an application's structure and implementation—including into any supporting functionality of which the application makes use.

1.1   Where Does a Design Team Begin?

Suppose you are designing an application. You've learned why your organization is creating it and know who the intended customers and users are. You've interviewed and observed some of those people to understand their activity and how the application will fit into it. You've gathered functional requirements from them and from Marketing. In short, you've performed a business-case analysis, a market analysis, a task analysis, created user profiles, and are ready to start designing the application. What's your next step?

1.1.1   Design (and Implement) the Functionality?

Some software developers start designing an application by designing the implementation of its functionality. They define the application's architecture—platform, functional modules, and communications protocols—and then immediately begin implementing the modules and protocols on the chosen platform. Only after the functionality is complete or nearly complete do they begin thinking about the application's user interface, that is, how people will operate the application.

This was the prevalent practice in the early days of software development, when engineers and scientists were still the main users of software applications. The design attitude was: "Get the functionality right, then slap a UI onto it. The users can adapt".

That approach never worked very well, even when the application users were highly trained scientists and engineers, but it became completely untenable when the use of computers and computer-based products and services began to spread to the general population in the mid-1980s.

However, the old "functionality first, then UI" underwent a resurgence in the early 2000s, as Agile software development practices became popular—a resurgence that continues to this day. Chapter 9 explains that this is based on an overly-simplistic but widespread view of Agile development, and that Agile development does not necessarily mean focusing on implementation before user interface. However, for now, suffice it to say that a significant number of software development teams today jump right into coding the functionality of an application before they figure out how people will operate it.

Fortunately, many software designers and developers know that the "functionality first, then UI" approach usually yields software applications that are hard to learn, hard to operate, and have questionable utility. Therefore, they reject that approach and instead begin designing an application by focusing on how people will use it.

1.1.2   Design (Sketch) the User Interface?

Unfortunately, many designers, especially those new to application design, believe that "focusing on how people should think about doing the task" means starting by sketching the content screens, control panels, and dialog boxes of their application. Such initial sketches—called *wireframes* in UI design jargon—are usually high-level and low-fidelity, showing only gross layout and organization.

However, starting by sketching the user interface amounts to starting with how the application *presents itself* to users. This tends to mislead designers into focusing on presentation issues, diverting them from designing what the presentations are supposed to present. Even if the resulting user interface is easy to use, the application may be useless.

Designing the presentation first also leaves designers with a single user interface. For example, personal-finance applications could be designed around records that look like checkbooks, making them hard to adapt for use in cars where access would be through voice.

1.1.3   Create a Conceptual Design

Instead, we believe that it is better to start by designing how people think about their tasks: the *concepts* users will think about when using the application, and how those concepts fit together into a structure. That is, it is best to begin the design by designing a *conceptual model*.

Designing a coherent, task-focused, conceptual model is the most important step in designing a user interface for an application. It has been called the "jewel in the crown" of user-interface design steps—the step with the largest impact on whether the application will make sense to its users [Card, 1993; Card, 1996].

Unfortunately, conceptual design is often skipped in software development. The result is a marketplace—and a Web—full of incoherent, overly-complex applications that expose irrelevant concepts to their users.

A key part of interaction design of an application is creating a *conceptual model* of that application. The purpose of conceptual design—of creating a conceptual model—is to get the concepts and their relationships right, to enable the user to create plans that accomplish the targeted tasks.

We believe that it makes sense to get the concepts and their relationships right before designing how those concepts will be presented or implemented.

Thus, start by designing *how the user would ideally think about the application* and how to use it to perform tasks. Indeed, shape the whole development process around creating and delivering a good conceptual model.

1.2   Conceptual Models

Let's consider some examples of conceptual models.

1.2.1   An Example: Calendar

We use a calendar as a repeated example throughout the book. To get started, here is a fragment of a conceptual model for a simple calendar:

> **domain**: events—meetings, appointments, registrations, travels.
> **purpose**: remember and managing events.
> **tasks**: schedule an event, reschedule an event, forget an event.
> **Structure**:
>> **objects**: event.
>> **operations**: add event, delete event.
>> **attributes**: date, time, location.
>> **relations**: subevent.

1.2.2   Contrasting Conceptual Models

The following are pairs of alternative and incompatible conceptual models. In each pair, either model could be acceptable depending on the requirements. However, the two models in each pair are incompatible, so the application designer must choose one of them (or invent another alternative).

Assume you are designing:

- A calendar for managing events. Is the calendar …

  (a) a list of dates, each with that date's events, or
  (b) a list of events, each with that event's date,
  (c) a set of sheets, each with a set of events, each of which has a date?

- An online discussion forum organized by topic. Is the structure…

  (a) a set of topics, each with postings and responses in one flat list, or
  (b) a hierarchy of postings, each potentially with responses?

- An application for creating newsletters. Is a newsletter ...

  (a) a list of items, or
  (b) a set of pages, each with layout of items?

- An application for ordering food. Does the application have ...

  (a) a list of restaurants with menus showing each restaurant's dishes, or
  (b) a list of dishes showing which restaurants offer it.

1.3   Important Decisions

Conceptual design decisions matter. Depending on how designers conceptualize an application, users of the application will think differently about using it: the objects will be different, the operations people can do on them will be different, and how people use the application will be different. Confronted with different ways of thinking (for example the contrasting pairs presented above), designers could try to avoid choosing, to have it both ways, but then users will get confused direction on how to think about the application and a confused understanding of how to use it. *Not choosing* is tempting, because these decisions are difficult to make; they usually require tradeoffs between simplicity and power and they always require understanding users' tasks. But whether or not designers create a conceptual model, in the end, the application will have one, even if it is just an unintentional—and often incoherent and unclear—result of the rest of the design process.

Conceptual design decisions are often tough, but they are essential. They are better made early, before they are blocked or locked-in by user interface design and implementation decisions. Get the bone structure right first, then flesh it out.

By carefully crafting an explicit conceptual model focused squarely on the target task domain, and then designing a user interface[2] from *that*, the resulting product or service will have a much better chance of being as simple as the task domain allows [Norman, 2010], coherent, and easy to learn. In contrast, designers who jump straight into laying out screens and dialog boxes, or choosing controls and widgets from a user-interface toolkit, are much more likely to develop a product or service that seems arbitrary, incoherent, overly complicated, and heavily laden with computer-isms and other irrelevant concepts.

Designers with strong backgrounds in human–computer interaction (HCI) and user-interface design are probably well-aware of the benefits of using conceptual models. However, our experience with our clients indicates that conceptual models of this

---

[2] Or multiple user interfaces to support different delivery devices.

sort are almost completely unknown outside of the HCI community, especially among web-designers and software programmers.

1.4   Conceptual Design's Place in the Application Development Process

A good conceptual model is central to a good product. A good development process will therefore keep the design of its conceptual model clearly in focus. This book therefore argues that conceptual models are central not only to the design of good products, but also to the processes of developing them.

Chapter 9 discusses in detail where and how conceptual models fit into the software development process, but for now, it suffices to say that conceptual design focuses the design process, coordinates design activities, and guides the design of the user interface, implementation, documentation, training, and support. Most design decisions impact or are impacted by the conceptual model. It is at the boundary between different perspectives, and a bridge between them.

1.5   The Benefits of Using Conceptual Models

Chapter 10 explains the benefits of using conceptual models, but to motivate you to read the rest of this book, we will summarize the main points here. The main benefits of conceptual models are:

- Produces a vocabulary,
- Facilitates creation of conceptual scenarios,
- Facilitates creation of user documentation, training, and support,
- Focuses user interface design: gives designers a clear target,
- Jump-starts and focuses the Implementation,
- Supports further task analysis,
- facilitates iteration of the design, including the conceptual model,
- Facilitates communication between team members,
- Saves time and money.

Since the design, the process, and the experience of use are all informed by the conceptual model, these all feed off each other and grow together.

1.6 Organization of Book

The book is organized as follows:

Chapters 1–2: Designing tools to help people perform tasks

1. **Introduction**
2. **Framework and Terminology**

Chapters 3–6: Conceptual models

3. **What Conceptual Models Are and Are Not**
4. **Components of a Conceptual Model**
5. **Representing Conceptual Models**
6. **Two Complete Examples**

Chapters 7–9: Applying conceptual models to the practical realities of design

7. **Essential Conceptual Modeling**
8. **Enhanced Conceptual Modeling**
9. **Process of Designing with Conceptual Models**

Chapter 10: Conclusions

10. **Benefits of Designing with Conceptual Models**

This chapter and Chap. 2 are intended to provide those new to the design of applications with sufficient background knowledge to understand the rest of the book. They set the context within which conceptual models are important. This chapter introduces the topic and the book. It offers "teasers" of topics presented later in the book and outlines the book's organization. Chapter 2 introduces the key ideas (e.g., people, domain, task, activity, application, user's mental model, conceptual model, user interface, implementation, and design process) and the terms used for them throughout this book.

Chapters 3 through 6 explain what conceptual models are. Chapter 3 introduces conceptual models, the purpose they serve, what they are, and what they are *not*. Chapter 4 describes how conceptual models are organized and structured. This will be the central chapter for most readers. Chapter 5 shows common ways of presenting objects/operations structures—the main component of conceptual models. Chapter 6 provides two complete, and so larger, examples.

Chapters 7 through 9 explain how to build and use conceptual models. Chapter 7 describes how common configurations of concepts (e.g., types, specialization) can be expressed using the objects/attributes/operations structure of conceptual models. Chapter 8 raises complex issues that may be important to address in some conceptual models. Chapter 9 describes how conceptual models fit into an iterative design process.

Chapter 10 enumerates and describes in detail the benefits that conceptual models can bring to the development of applications, both for designers and for people using the applications.

Framework and Terminology

2

People carry out activities in support of doing tasks in domains of human endeavor. These days, they often use interactive applications in support of those activities.

Developing interactive applications involves a large and complex set of theories and practices. This book addresses only a single aspect of that: we contend that conceptual models (CMs) of the applications should be a core part of the work of developing interactive applications.

This chapter establishes a framework for interactive applications as a context for the rest of the book. In doing so, it presents ideas and introduces the terms used throughout the book. This framework addresses both the use of interactive applications and the development process within which conceptual models should be central (Fig. 2.1).

2.1 Framework for Interactive Applications and Their Use

At the center of this framework are people doing tasks using interactive applications.

2.1.1 People and Domains

In their work and their play, people are engaged in domains of human concern. They write letters, send messages, play music. They manage their photographs, their finances, their time. They write books, give presentations, create personal history albums. They attend meetings, converse. They check the weather, plan for travel.

© The Author(s), under exclusive license to Springer Nature Switzerland AG 2024
J. Johnson and A. Henderson, *Conceptual Models*, Synthesis Lectures
on Human-Centered Informatics, https://doi.org/10.1007/978-3-031-50852-3_2

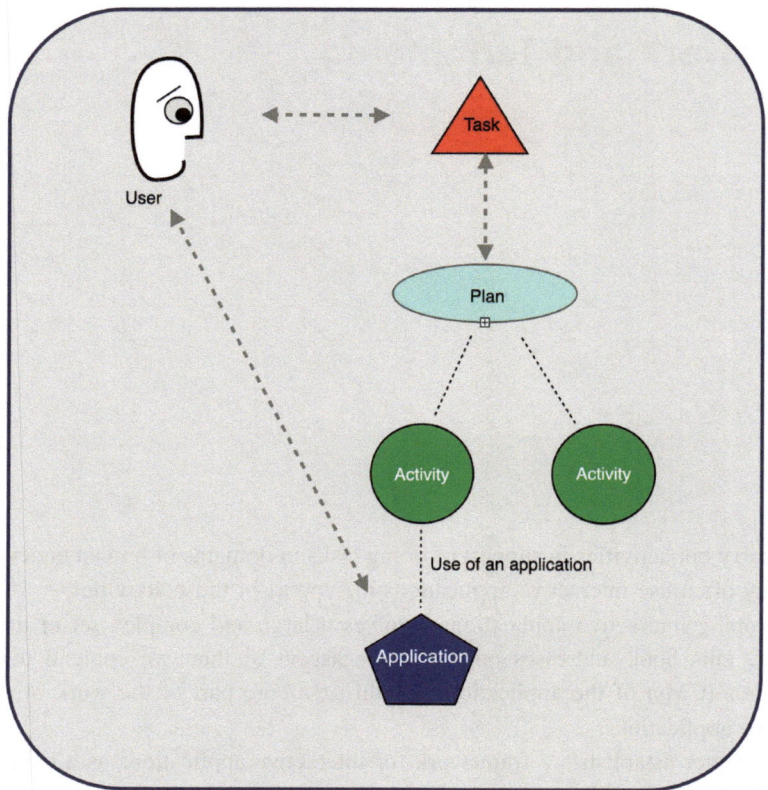

Fig. 2.1 Application in use by a user with a plan to perform a task in a domain

For example, the domain of banking transactions could include making deposits, reviewing balances, making withdrawals. A broader (more encompassing) domain might include all of these, together with bill paying, transferring money between accounts, and budgeting.

People: Mary and George.
Domain: Managing household calendar of events.

A person can be engaged in multiple domains at the same time.

2.1.2 Tasks

Within domains, people perform tasks. A task is a well-defined chunk of activity that has a goal that can be achieved with activities, some of which can be performed using applications.

The tasks within a domain are sometimes simple, but more often they are rich and complex. They evolve as time passes. They can overlap each other, with one activity serving multiple purposes. Sometimes they finish; sometimes they continue indefinitely.

Task 1: Mary agrees to meet with Susan.
Task 2: George adjusts the timing of his lunch with Don.

2.1.3 Plans and Activities

A person develops a *plan* of activities for doing a task. Sometimes the plan is explicit; often it is only implicit. Sometimes it is developed before any activity begins; often it emerges as the activity proceeds.

There can be different plans for carrying out any task.

Plan 1: Mary plans to record her meeting with her iPhone.
Plan 2: George plans to remember the change in the lunch schedule.

2.1.4 Applications and Uses

From the perspective of people with tasks to do, technology is of no interest by itself. However, of interest to us in this book are activities that are supported by interactive technology. We refer to interactive technologies as "applications".

People make *use* of applications as activities in their plans.

Application: Mary uses the Calendar application to manage her events.
Use: Mary adds the meeting with Susan to her online calendar.
(Non-interactive activity: George remembers the new time for his lunch.)

Tasks may be accomplished in a variety of ways. For example, a person can sell stock through a website, through an automatic phone tree, or through a broker. Therefore, at some point, a person must learn what applications are available and how each can be useful for the activities that they want to carry out. The person must make the connection between tasks and applications. Long ago, Tom Moran called it the External-Task/Internal-Tasks mapping (ETIT) (Moran 1983).

An effective application can have a strong influence on how people conceptualize their tasks, both because people can easily create plans that make use of the application, and because the capabilities of the application can affect how people view their tasks. For example:

> climbing a tree is easier if you wear tree-climbing spurs, but if you have a ladder, the need to climb a tree (e.g., to pick its fruit) disappears;
> the availability of spreadsheet applications not only leads people to recast computations as spreadsheets, but also to explore alternatives, which is much easier with a spreadsheet than a calculator.

Applications thus can also become "tools for thought". Richard Young in "The Machine within the Machine" has described how tools reshape not only work, but also how people think about work (Young 1981).

2.1.5 Conceptual Models

In Chap. 4, we present conceptual models in detail, including their components. Here, we introduce the terms that the rest of this book uses for discussing conceptual models.

A conceptual model for an application is composed of the following components:

- domain—*including common tasks*
- *purpose and goals—including tasks to be supported by the application*
- vocabulary—the *accepted terms for describing the conceptual model*
- *structure*—a collection of *concepts*, of four sorts: *objects, operations, relations,* and *attributes, and their interdependencies*
- conceptual *scenarios*—showing how the concepts support the performance of the tasks
- resolved and *unresolved issues.*

In Chap. 5, we discuss various ways to present conceptual models.

In Chap. 6, we give some more complete examples.

2.2 Beyond Conceptual Models

To be used, applications must be accessed and implemented. Therefore, the framework includes two additional aspects: user interfaces, and implementations (Fig. 2.2).

Fig. 2.2 User's mental model of application in use

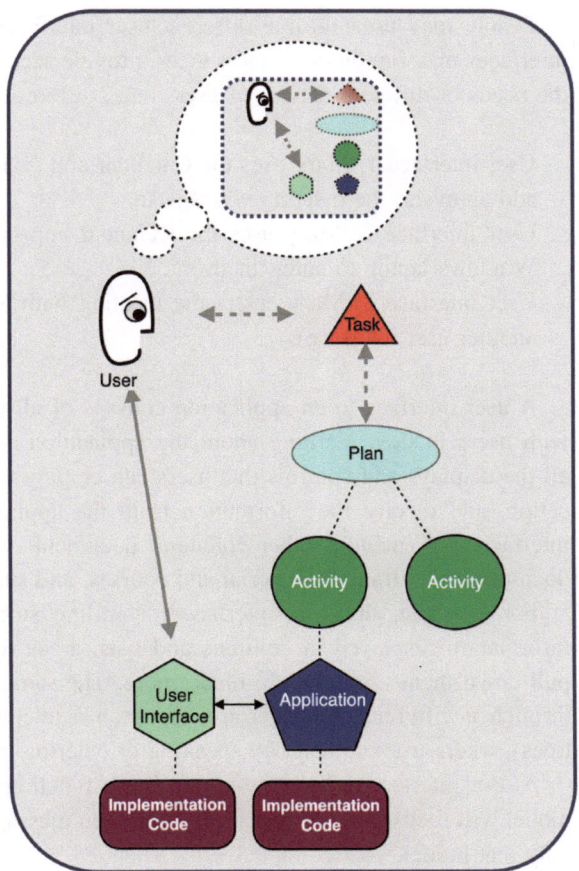

We contend that user interfaces and implementations of an application should be based upon a conceptual model of the application. They therefore depend upon the conceptual model. However, the conceptual model should not depend upon them. Therefore, although this book discusses user interfaces and implementation in much less detail, we include them here to support those discussions.

2.2.1 User Interfaces

While people can *think* about using an application at the functional level, to *take action* using the application they must access its functionality through an interactional mechanism. That is, through a user interface (UI).

People may use multiple different user interfaces to access an application; the user interfaces may run on different devices, provide access to different capabilities, or address the needs of different classes of users (e.g., different ages, abilities, or circumstances).

User interface 1: Mary uses the Calendar app (via the user interface on her iPhone) to add an event: the meeting with Susan.
User interface 2: Mary uses the Calendar app (via a desktop user interface) on her Windows laptop to enter the event.
User interface 3: Mary enters the meeting with Susan through her Apple watch (yet another user interface).

A user interface to an application consists of all means and mechanisms that interact with users in their learning about the application and making use of it. These include all the displays and controls that users can employ to invoke the operations of the application, and receive the information from the application indicating its state. The user interface also includes other collateral documents and services, such as help material, documentation, training material and courses, and support services.

For example, the user interface for selling stock through a website may involve information displayed in columns and lists, areas for entering text, links and clicking, pull-down menus, and much, much more. The same application may also be accessible through a different user interface based on a telephone and multi-level menus (phone trees), where users interact by speaking or entering numbers.

A user interface delivers an application's functionality to its users. A clear conceptual model will focus the user interface to stay "on message," presenting all of that model and only that model.

2.2.2 Implementation

Both applications and user interfaces can be implemented with software.

Implementation of the application: The Calendar application runs in part in the cloud and is coded in C++.
Implementation of the user interface: Mary's watch is built by Apple and is coded in Swift.

An application's users do not need to have, and indeed rarely do have, any idea about how the application is implemented.

2.3   Thinking About Applications

This raises the central question that motivates this book: How do people think about applications? What is the information that a person needs concerning what an application does, so that they can figure out whether that application will help them with their task? Correspondingly, how is that information expressed: how are applications described?

2.3.1   User's Mental Model

To use an application, a person must have some understanding of it. Such an understanding is called a "mental model" because it is in the person's mind. The mental model is usually incomplete—even somewhat incorrect.

A person will also have an understanding of the task that they are trying to accomplish. This too may be partial and/or incorrect (Fig. 2.3).

Everyone has their own mental models of their tasks and applications. For example, although different people use the same copier, they may understand it differently. They

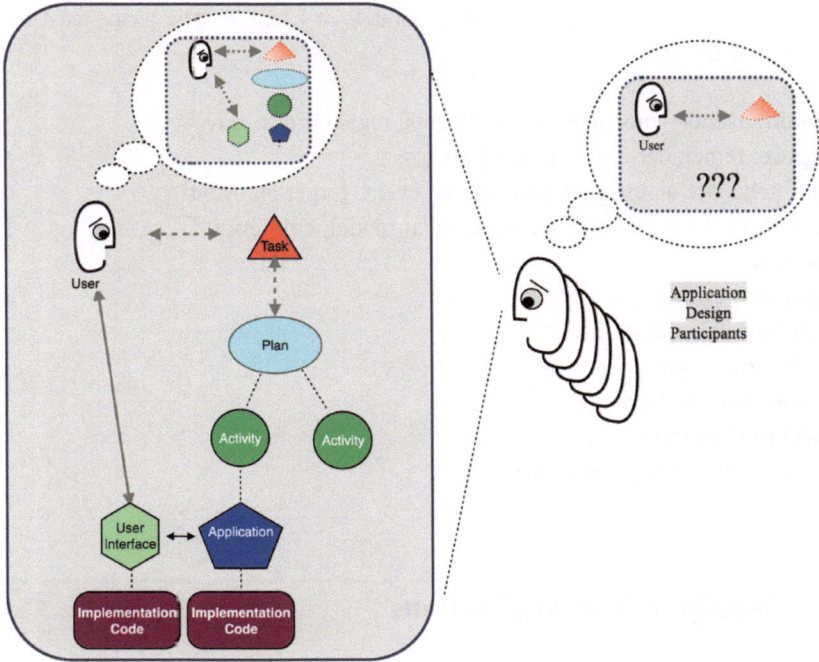

Fig. 2.3 Designer's conceptual model

may understand some aspects well, and others less well or not at all. Also, a person's mental model may well change as a result of experience with the application.

Thus, mental models are personal, partial, uncertain, and dynamic. They are also usually not directly accessible to designers. Research has even shown that an application user's mental model is often not entirely available to the user (Card 1996)!

2.3.2   Designers' Conceptual Model

One important goal in designing applications and their user interfaces is to help the application's users develop mental models that correspond to the application. The key to achieving this is that the designers have designed a good and clear model of the application. This is the designers' conceptual model. It is the *ideal user's mental model*—the model that the designers strive to have their applications and their user interfaces induce in their intended users.

2.3.3   Terms for Conceptual Modeling

We discuss the components of conceptual models in Chap. 4. The terms used to discuss those components are:

domain: events—meetings, appointments, registrations, travels
purpose: remember and managing events
tasks: schedule an event, reschedule an event, forget an event
vocabulary: terms used in the conceptual model, e.g., event
structure:
    objects: event
    operations: add event
    attributes: type
    relations: subevent
conceptual scenarios
resolved and unresolved issues.

2.4   Descriptions of Applications

There are many ways to describe applications, serving the needs of the various participants in the application design and development.

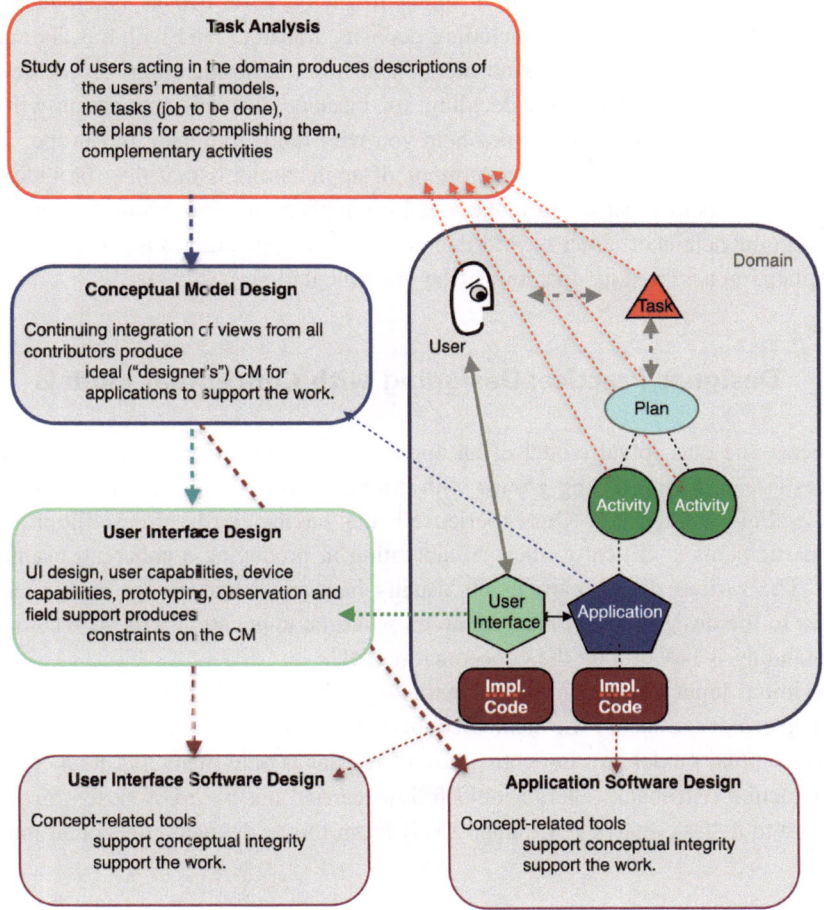

Fig. 2.4 Design of a conceptual model

Marketing wants to ensure that the application helps the target users perform their tasks (task-based description).

Documenters and users want to describe how to use menus and touch buttons to get the task done (user interface based description).

The software implementers want to describe their code (implementation based description).

However, all these participants must agree on, and be driven by, how users are going to think about using the application. Therefore:

Everyone needs to agree on a conceptual model (model-based description).

For example, an application for banking might be described as manipulating bank accounts, enabling transactions (including deposits, transfers, and withdrawals), and providing reports. Note that there is no mention of how these operations are achieved, but there is adequate information for deciding, for instance, that this application will indeed help you make a deposit and will **not** help you wire funds to a bank in Europe.

The advantage of models as descriptions of applications is that they provide enough information to help a person to make a task-to-application plan, without overwhelming them with the details of either how to interact with the application (user interface) or how the application achieves its operations (the implementation).

2.5   Designer Practice: Designing with Conceptual Models

In general, the conceptual model of an application emerges from a design practice that involves a team of interacting groups with different skills, responsibilities, and perspectives (see Figs. 2.4 and 2.5). Our experience is that having clarity of description available to all participants is critical to their collaboration in producing a coherent, usable application. The various design participants usually have good reasons for wanting to make changes to the design, but all must agree on what the application will be. That common understanding is based upon the conceptual model.

The initial input might come from marketing: "We need an application that lets regular people manage their own bank accounts", with suggestions as to the key concepts for a conceptual model. An in-depth study of regular people managing finances, including interaction with bank tellers should follow, carried out by a UX design group. This must lead to a "task analysis" [Young, 2008] From this, a detailed conceptual model can

Fig. 2.5 Design and design team coordinated by conceptual model

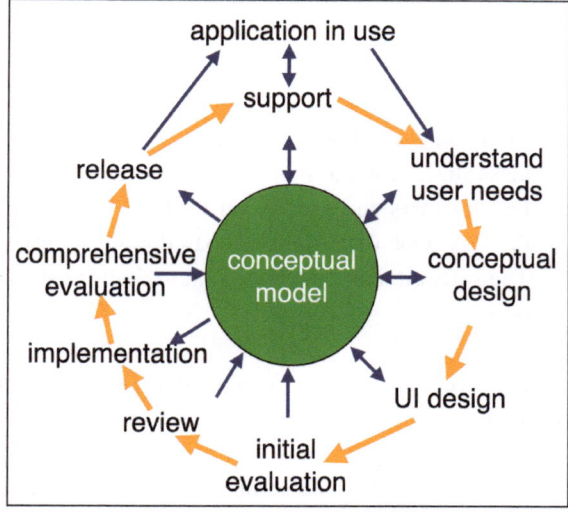

be produced. The others on the team can react to it, suggesting changes, schedule, and budget. The discussion continues until the conceptual model stabilizes enough to make decisions about development including assigning resources to it.

At this point, the key perspectives on the design will have been considered: task analysis and design, application (conceptual model) design, user interface design, software design, marketing, support, and so on.

As the work proceeds, surprises will arise, which will prompt further discussion and yield a deeper understanding and clarification of what is required. Assuming all goes well, a product will be produced, sold, shipped, and launched, at which point the application's intended users will again have their say—remember that they were engaged in the task analysis—and the application will continue to evolve. Indeed, as the domain changes, the conceptual model should change with it, evolving over the life of the application, acting as a base for clear discussions and negotiation, and keeping everyone aligned.

We discuss the role of conceptual models in the design process further in Chap. 9.

What Conceptual Models Are and Are Not 3

Before we explain *how* to design conceptual models of interactive applications, we must explain *what* conceptual models are—and are *not*. That is the main purpose of this chapter.

However, before we explain that, we must first step back and explain what models are and what it means to create a model of something.

3.1 What are Models?

Children often build and play with models: model airplanes, model cars and trucks, model trains, model kitchens, model animals, and even model people (called "dolls"). Engineers and architects preparing to build something often build one or more models of the thing first: bridges, buildings, landscapes, freeways, monuments, spacecraft, power grids, etc. Scientists researching natural phenomena create models to help them predict events: ocean currents, weather patterns, earthquakes, etc.

These widely different models have one thing in common: they recreate *some*—but not all—aspects of the modeled object or phenomenon. For example, model airplanes, trains, and bridges recreate the shape and appearance of real airplanes, trains, and bridges, but not their size and weight. If models recreated *all* aspects of the modeled thing, they wouldn't be models; they would be the real thing.

To make a model of something, one must decide *which* aspects of the thing to recreate—i.e., to model—and which aspects *not* to recreate. What to model and what not to model depends on the *purpose* of the model. Depending on how the model will be used, some aspects of the real object will be important to replicate in the model, and other

© The Author(s), under exclusive license to Springer Nature Switzerland AG 2024
J. Johnson and A. Henderson, *Conceptual Models*, Synthesis Lectures
on Human-Centered Informatics, https://doi.org/10.1007/978-3-031-50852-3_3

aspects won't be. Engineers creating a model of a bridge often want the model to simulate stress-handling characteristics of the real bridge, but don't want to recreate the size
and don't care about the color.

3.2 What Are *Conceptual* Models?

A conceptual model is a model of an interactive application. It embodies aspects of the
real application that are important for communicating to the application's design team and
users a high-level understanding of how the application functions and the tasks it can help
its users perform. A conceptual model ignores aspects of the application that do not serve
that purpose, such as user interface details, implementation details, marketing channels,
and price.

Because a conceptual model is a very high-level specification of the application's
functionality, it can be regarded as a focused *description* of the application. Interactive
applications can be described based on: their detailed user interface, their implementation,
their supported tasks, and the concepts (objects, attributes, and operations) they embody.

3.2.1 Unsatisfactory Descriptions of Applications

Describing an application in terms of its *user interface*[1] components is problematic
because applications can have more than one user interface. For example, an application might have one user interface for normally sighted people using desktop and laptop
computers, another for sight-impaired users, another for smartphones, and another for telephone menu-trees and smart speakers. The details of the possible user interfaces could be
quite different, leading to quite divergent descriptions of the same application.

Describing an application in terms of how it is *implemented* might satisfy software
developers but would be much too detailed and subject to change. Such a description
would also be incomprehensible to non-programmers on the team and to the application's
end-users.

3.2.2 Better: Task-Level Descriptions

A slightly better way to describe an application is to map the *tasks* it supports to the steps
required to accomplish those tasks. Such a description would help people—team members
and end-users—understand how to navigate their way through tasks that are supported and
documented. However, if the application supports many tasks, a task-based description
may be overwhelming. Also, a task-based description won't help people figure out how

---

[1] Sometimes called the *front end*.

to do undocumented or unanticipated tasks, troubleshoot problems, or predict which tasks it does *not* support. Nonetheless, task-level descriptions are useful, so designers often use them to augment conceptual models. Task-level descriptions exist in many forms, including *task models*, task scenarios, and *use-cases*. We discuss these in more detail later in this chapter.

3.2.3   Best: Object-Operations Model Descriptions

The best way to describe an application is in terms of the high-level concepts it embodies: the types of content *objects* the application's users can manipulate, the attributes of each type of object (which comprise the object's state), the *operations* on each type of object, and the *relationships*—functional and otherwise—between objects. People can internalize such a *conceptual* model to form a *mental* model of the application, allowing them to predict what tasks they can do with it, how to do those tasks, and what tasks the application does not support.

For example, a banking application might be described as manipulating *bank accounts*, enabling *transactions* (e.g., deposits, transfers, and withdrawals) and generating *reports* (e.g., current balance). Without mentioning a specific user interface, this description (or model) provides enough information to decide, for instance, that this application will help you make a deposit and that it will **not** help you apply for a bank loan to buy a home.

This objects/operations model-based way of describing applications is the focus of this book.

3.2.4   Our Approach to Conceptual Modeling

As is mentioned in Chap. 1 and described in detail in the Appendix, conceptual models have been invented independently by researchers and practitioners in several different sub-fields of computer science and information technology. Therefore, it is not surprising that the different approaches vary in their goals and which aspects of applications they model. For example, some software development experts want a conceptual model to specify the *entire* application so the implementation code can be generated automatically. In contrast, UI/UX designers who advocate conceptual modeling tend to be less ambitious; they just want to model the user-facing concepts of an application or task domain to help focus their design on the target task domain and to clarify, organize, and simplify it. We—this book's authors—are designers, so the latter is closer to our view of conceptual modeling. In the rest of this chapter and in the rest of the book's chapters, we describe our approach to conceptual modeling.

3.2.5   High-Level Description of an Application

As we explained above, a *conceptual model* is a high-level description of an application. Without reference to a user interface or an implementation, it enumerates concepts— objects, attributes, and operations—in the application that users can manipulate, describes how those concepts relate to each other, and explains how those concepts fit into tasks that users perform with the application. In so doing, the conceptual model provides people with an "intuitive" grasp of what they can do with the application and what concepts they must understand in order to use it successfully.

More specifically, a conceptual model specifies and describes:

- the target task domain, purpose, and high-level functionality of the application.
- the **concepts** the application exposes to users, including the task-domain data objects that users create and manipulate, their user-visible names, their attributes (options and settings), and the operations that can be performed on them.
- the **relationships** between these concepts.
- the **mapping** of task-domain concepts to application concepts, usually expressed as task scenarios written in terms of the conceptual model.

As a simple example, consider an alarm clock. A conceptual model for it might state:

- The clock stores the current **time of day**, continually updating it to track the passage of time.
- It displays the current time constantly.
- Users can set the current time.
- Users can set an **alarm** at a specified time, or no alarm.
- When an alarm is set and the current time equals the set alarm time, the alarm is triggered.

Our model of a simple alarm clock allows only one alarm to be set at a time. It does not include concepts of *multiple alarms*, possibly with different *alarm-indicators* (e.g., sounds, flashing lights, radio ON, start music-player) for each alarm. It *could* include such concepts if the manufacturer wanted to offer a fancy alarm clock and could justify the extra complexity and cost. However, if support for multiple alarms is not a requirement, such concepts would be better left out of the conceptual model to keep it simple.

Similarly, the manufacturer might want the clock to indicate the **current date** and **day of the week** as well as the **time of day**. If so, the conceptual model must include those concepts. However, features should not be added lightly because they often increase the number of—and interactions between—concepts in the conceptual model and hence its complexity, which makes the application harder for people to learn and use.

The alarm clock could be made even simpler. Like mobile phones, it could obtain the time from an external source, such as a cellular link, GPS, or wi-fi Internet. If the external connection and querying of the time were totally automatic and error-free, the conceptual model for such an alarm clock could be very simple indeed: it would not need to store the time locally, update it constantly, or provide a way for people to set the time.

However, suppose getting the time from an external source was *not* fully automatic and error-free? Including concepts for managing that—e.g., detecting external time-sources and connecting to them—could easily make the clock *more* complicated than an ordinary alarm clock.

Sidebar 1: A More Complex Example

Consider an online catalog for a city library. The conceptual model might include:

- **High-level functionality and purpose**: e.g., the application is intended to support finding items in the library, replacing a physical card-catalog.
- **Concepts**: e.g., **catalog** (attributes: items; operation: search), **item** (attributes: title, ISBN, status; operations: check-out, check-in, reserve), **subtypes of item** (e.g., book, periodical issue, LP, video), **user account** (attributes: name, items checked out), **librarian** (attributes: name, desk location, phone).
- **Relationships**: e.g., a book is one *type* of item, periodical volumes *contain* issues.
- **Mappings**: e.g., each item in the system corresponds to a physical item in the library.

3.2.6   Basis for Users' Mental Model of the Application

A conceptual model describes how designers want users to think about the application. In using an interactive product or service, reading its documentation, and talking with other people who use it, people develop an understanding—called a *mental model*—of how the product works (Card 1996; Young 2008). This allows them to predict its behavior and generalize what they learn to new situations. Developing this understanding is an important part of learning to use a software application.

Ideally, users' understanding of the application should match what the designers intended. Otherwise, users will often be baffled by what it is doing. If the designers explicitly design a conceptual model and then base the detailed user interface design on that, chances are greater that users' and designers' understanding of the product will match.

3.2.7   Design Goals for Conceptual Models

The goal is to keep a product's conceptual model: (1) as simple as possible, with as few concepts as are needed to provide the required functionality, and (2) as focused on the users' tasks as possible, that is, with concepts that map as directly as possible onto the concepts of the task-domain and are conducive to making plans for doing tasks.

Simple

A conceptual model should be as simple as possible while providing the required functionality. An important guideline for designing a conceptual model is: "Less is more".

For example, in a Web search function, do the intended users need boolean search capability? Would they use it if provided? If not—if a simple keyword search function is enough—designers shouldn't complicate the conceptual model by adding boolean search capability.

Similarly, suppose a To-Do list application needs the ability to assign a priority to a To-Do item. If user-research indicates that users want two priority levels (low and high), designers should not complicate the conceptual model by generalizing the requirement to include more priority levels (e.g., 1–10) (Beyer and Holtzblatt 1997).

Beware: attaining simplicity is not simple! Much thinking, testing, and rethinking are often required to discover a conceptual model that is simple enough but not too simple to provide the required functionality.

Task-Domain-Focused

Conceptual models should map as directly to the target task domain as possible. This reduces users' difficulty in translating concepts between those of their task and those of the application.

When the operations a tool provides don't match the operations of the task domain in which a person is working, the person must figure out how to combine the tool's operations to comprise or approximate the task-domain operations. "Hmm. I want to accomplish X, but this tool only lets me do A, B, and C. How can I put A, B, and C together to get X?". This imposes on the person a cognitively taxing *problem-solving* task in addition to the task they wanted to accomplish; they must form a *plan* and then execute that plan. Norman (Norman and Draper 1986) calls this difficulty of translating between the tool and the task domain "the gulf of execution".

Contrast inexpensive photocopiers intended for use in homes and small businesses with large, expensive, feature-laden copiers intended for intensive use in print-shops and large businesses. When a person uses a simple copier to do a complex copying job that the copier doesn't support directly, the person must first devise a more complex plan of operation: a sequence of steps using operations the copier *does* support together with other operations provided by the person themselves or by other tools. For example, if the goal is to make a double-sided copy from a single-sided original document, the plan might be: (a) separate the original into odd and even pages, (b) copy the odd pages, (c) put the copies back into the paper-hopper, (d) copy the original's even pages, (e) collate the original and copies to put the pages in proper sequence. Then the person must *execute* that plan carefully. The need to devise a plan to bridge between the desired goal and the tool's available operations is the gulf of execution. Using a larger, expensive copier doesn't have that gulf; they provide most of the higher-level operations that users need, directly.

As the copier example makes clear, there might be a very direct mapping from one task to the conceptual model of an application, but the mapping from *another* task to the same conceptual model might be more complex. The conceptual model determines how much of a task domain can be served by the application through a relatively direct mapping, and how much will require more elaborate plans. Better conceptual models will map to large task domains with simpler plans. Sidebar 2 provides an example of applying the task-focus design goal in a conceptual model for a complex application.

Sidebar 2: Redesigning a Conceptual Model for Analyzing Protein Interaction

One of us advised a company that was designing software to control an instrument for biologists to use to analyze protein interactions. An early prototype of the software was designed based on concepts about controlling the instrument, e.g., move sample-extraction pipettes 10 cm rightward, lower pipettes 5 cm (into protein sample well), activate suction, stop suction, raise pipettes 5 cm, move pipettes 20 cm leftward, lower pipettes 3 cm (to testing plate), discharge 0.01 ml of protein sample A on to testing plate, wait 30 s, activate protein-imaging lasers, etc.

Biologists who were shown this prototype rejected it completely. A typical comment was: "I don't want to control this machine; I want to analyze protein interactions." The conceptual model was re-designed based on protein-interaction analysis concepts. Instead of specifying a sequence of machine actions that would cause a protein to be analyzed, biologists specify a protein analysis *protocol*, e.g., interact three different concentrations of *ligand*-X with five different concentrations of *analyte*-Y. When a protocol is executed, the instrument translates it into the required sequence of machine actions. The new conceptual model matched how biologists think about analyzing proteins.

3.2.8   Summary: What a Conceptual Model is

Conceptual models are best thought of as *design-aids*—a way for designers to straighten out and simplify the design and match it to the users' task domain, thereby making it clearer to users how *they* should think about the application. The designers' responsibility is to devise a conceptual model that seems natural to users based on the users' familiarity with the task domain. If designers do their job well, the conceptual model will be the *basis* for users' mental models of the application (see "Not the Mental Model", below).

A conceptual model of an interactive application is, in summary:

- the *structure* of the application: the objects and their attributes, operations, and relationships,
- an *idealized, high-level, and often simplified view* of the application's functionality—the model designers hope users will adopt as their *mental* models.

3.2.9   Close Relative: Information Architecture

Web designers and information architects often include a stage in development that is similar to conceptual design: information architecture. Designing an information architecture for a website involves deciding how to organize the information it will present. This yields site *structure* and content *categorization*.[2] The goal of information architecture is:

- to organize content to help website users find what they are seeking, and
- to promote the goals of the website owner.

Information architecture is like conceptual design in that it is concerned with conceptual *organization* and *structure*, rather than with presentation. Information architecture also is similar to conceptual design in that it normally concerns only *user-facing* concepts.[3]

Nowadays few websites are just static pages of information. Most offer products or services to purchase; posts to view, react to, and comment on; search capabilities, online document preparation, etc. Such websites are actually online applications, not so different from the applications that come with or that people install on digital devices. For Web-based applications, the conceptual model includes the information architecture, but goes beyond it to include the objects and operations comprising the site's functionality.

---

[2] Sometimes called "ontologies".

[3] Some people include implementation concerns in IA, e.g., whether an underlying database is relational, and if so, how it is represented in tables.

3.3   What Conceptual Models Are *Not*

To further clarify what conceptual models *are*, let's consider some things that they are *not*.

3.3.1   Not Users' Mental Model

The conceptual model is *not the users' mental model* of the application. As is mentioned in Chap. 2, users of an application form mental models of it to allow them to predict its behavior. A mental model is the user's high-level understanding of how the application works; it allows the user to predict what the application will do in response to various user actions. Ideally, a user's mental model of an application should be *like* the designers' conceptual model, but in practice, the two models may differ. Even if a user's mental model is the *same* as the designer's conceptual model, they are distinct models.

Users' mental models can include all sorts of extraneous concepts and relationships. For example, a person's mental model of their watch might include the fact that it tends to gain time and so must be reset every week or so. It might also include the knowledge that the alarm is not loud enough to wake them, and that the watch is a family heirloom that should under no circumstances be lost or damaged.

Second, users' mental models, especially of complex software-based applications, are difficult for designers to pin down. Mental models cannot be observed directly and users can rarely describe them accurately. Designers usually must figure out users' mental models by observing people's behavior—including successes, difficulties, terminology, and questions—and asking them to explain their actions and expectations.

Finally, users' mental models are only an approximation of the designers' conceptual model. Users form mental models of an application by observing how it behaves, reading its documentation, and talking with other users about it—in short, through their *experience* with the application. As a user's experience with an application grows, the user's mental model changes.

3.3.2   Not a Design Metaphor

Conceptual models are not the same thing as design metaphors. Many software user interfaces are based on metaphors[4] with common objects, often from the physical world. This leverages people's (presumed) advanced knowledge of those objects. For example:

- Some computer operating systems use a "desktop" metaphor (Johnson et al. 1989).

---

[4] They are more properly called "design analogies", but the term "metaphor" has stuck, so we are stuck with it.

- Most eReaders use a "book" metaphor.
- Most computer calculation functions use a "pocket calculator" metaphor (Johnson 1985).
- Many Delete functions use a "trash can" or "recycle bin" metaphor.

Designing software based on analogies with familiar physical objects has disadvantages as well as advantages, such as suggesting to users that all the laws and limits of the physical world apply to the software as well, when in fact they usually don't (Halasz and Moran 1982; Johnson 1987).

One or more design metaphors may be *included in* a conceptual model, but a conceptual model includes much besides the design metaphor, as described above and in Chap. 4.

3.3.3   Not Just Task Models, Task-Level Scenarios, or Use Cases

As mentioned earlier in this chapter, designers often create task-based descriptions of an application's functionality. The purpose is to ensure that the design accounts for all tasks that the application is supposed to support.

One type of task-based description is called a task model (Beyer and Holtzblatt 2016). There are many different types of task models (Bowen et al. 2021). We describe and illustrate some of them in Chap. 9.

Task-level scenarios and *use-cases* are another task-based way to describe applications. They are brief stories about the tasks that users will perform, mapping users' goals to high-level application functionality. Software engineering experts define a *use-case* as: "a description of steps or actions between a user (or 'actor') and a software system which leads the user toward something useful" (Bittner and Spence 2003). Both task-level scenarios and use-cases are expressed in a design-neutral way, so as not to specify a user interface. For present purposes, task-level scenarios and use-cases are similar enough that we will lump them together under the term "task-level scenario".

A task-level scenario for a hotel-room telephone might be: *room-guest returns to room and checks to see if anyone has left voice-mail.* Similarly, a task-level scenario for an online banking application might be: *customer transfers funds from savings account to checking account.*

Task models and task-level scenarios emerge from study and analysis of the task domain—through interviews, ethnographies, focus groups, contextual inquiry, and other methods. They can either be *input* to the design of the conceptual model or they can *emerge* from it. Therefore, task models and task-level scenarios are often included as a component of conceptual models (see Chap. 4) and sometimes used to present conceptual objects, attributes, and operations (see Chap. 5) to non-technical stakeholders. However, task models and task-level scenarios do not by themselves constitute a conceptual model.

3.3.4   Not Based on the User Interface

The conceptual model of an interactive system is *not based on a design for the user interface*. It is not about how the application looks, feels, or sounds. It does not mention clicking, tapping, or swiping, keystrokes and mouse-actions, screen graphics and layout, commands, navigation schemes, dialog boxes, buttons, menus or other controls, data presentation, links, pages, screens, or error messages. It does not say whether the application is operated through a GUI on a personal computer or by voice commands over a telephone. It does not mention whether the application presents text in English, Spanish, Chinese, German, or any other specific human language.

For example, the clock conceptual model described above says nothing about whether the clock is set by voice commands, buttons on the top, or knobs on the back, whether it has rotating hands or a numerical display, or whether the alarm indicator is a sound or a blinking light. Those are *user interface* details that are not important until the basic concepts and structure of the design have been determined. The conceptual model only specifies that the clock presents the current time, the time can be set, and the clock can raise an alarm.

Our conceptual model also says nothing about how the clock looks. Digital clocks (numbers) and analog clocks (hands) vary in their presentation. There is nothing conceptual about this difference at all: at the level of the conceptual model time is time, and tracking time and making it available, however presented, is what clocks do.

A common mistake made by designers who are new to conceptual modeling is to equate the main screens or pages of an application or a website with objects in the conceptual model. Application screens and website landing pages are parts of the *user interface*; they are not data objects created and manipulated by people using the applications. They may *present* conceptual objects (e.g., posts, comments, bank accounts, available flights, songs), but the pages themselves are not conceptual objects. This is discussed in more detail in Chap. 7.

User-experience and user-interface designers often create prototypes of applications they are designing, using prototyping tools such as Figma, Adobe XD, JustInMind, Proto.io, and others. Such tools allow designers to create prototypes that simulate the planned application's appearance and even—in limited ways—its behavior. These are indeed models of the application, since they satisfy our definition (see above) of what a model is. However, they are not *conceptual* models. They are models of possible user interfaces for the application.

3.3.5   Not Formal Specifications to Automatically Generate or Simulate Applications

Some computer science researchers and software engineering experts assert that a conceptual model of an application should specify *everything* about it: conceptual objects, attributes, and operations; data structure; information flow; operational flow and constraints; user interface; and implementation architecture. The goal of this view of conceptual modeling is to allow automatic generation of the application or at least an operational prototype of it.

To achieve that goal requires specifying conceptual models in formal, machine-executable languages. There is no formal language that can specify all of the above-mentioned aspects of an application's design, so each aspect of the design would be specified using a different formalism. Examples of such formalisms include: entity-relationship charts for the data structure and information flow, UML diagrams for the implementation architecture and possibly the Objects/Operations structure, and predicate logic for the operational flow and constraints.

We do not consider automatic generation of applications from conceptual models to be practical or desirable, mostly because we do not expect software designers and developers to be facile enough with all of the formal languages required to specify every aspect of an application. For example, formal logic is **at least** as hard to learn and error prone as coding in conventional programming languages.

3.3.6   Not the Implementation Architecture or an Implementation-Based Description

A conceptual model is *not an implementation architecture.* An application's implementation architecture contains technical objects (possibly including classes, attributes, methods, and control structures) that are required to implement the application. In contrast, a conceptual model includes only concepts required by the task domain.

Although conceptual models and implementation architectures for an application are distinct, they can be related. A conceptual model often shapes the implementation architecture: some classes, variables, and methods in the implementation architecture typically correspond to concepts (objects, attributes, and operations) in the conceptual model. For example, a BankAccount class in the implementation architecture underlies the concept of a bank account object in the conceptual model. However, an implementation architecture also includes implementation mechanisms that are of no concern to users and thus have no place in the conceptual model. It is unfortunate that the terminology of conceptual design and object-oriented software design collide. Designers must take care when working at the intersection of these perspectives.

Both the implementation architecture and the conceptual model *describe* the application, but they describe it in different terms, for different purposes. The conceptual model describes the application in terms of concepts that are relevant to tasks to be supported by the application. The implementation architecture describes the application in terms of the components, modules, and layers of which it is constructed. Most of the technical objects in the implementation are of no concern to users, and therefore should not be exposed to them.

Our clock conceptual model, for example, says nothing about the internal workings of the clock, e.g., if it is analog or digital, if it has a quartz crystal and step-down gears, or even if it is electrical or mechanical. Those are *implementation issues* that are irrelevant to using the clock. There is nothing conceptual about the difference between digital and analog where clocks are concerned: users think of them as being the same—they just tell time.

Pure implementation objects, such as the *database* underlying a personnel management application, are of no interest to the application's users and so should not be exposed to them. Therefore, such concepts have *no* place in the application's conceptual model. Including them in a conceptual model detracts from the task-focus and reduces the learnability and usability of the product, no matter how much effort goes into designing the user interface. A clock, for example, might have a *time register* where the current time is stored, but that is of no interest to the clock's users, so a conceptual model would exclude that concept.

3.3.7   Not Just the Information or Content Architecture

Conceptual models as described in this book specify not only the structure of the information in an application (often referred to as its *content ontology*), but also the available *operations* on that information, that is, what the application's users can *do* with the information.

3.3.8   Not Product Designer's "Concept Design"

Finally, to avoid any confusion, the terms *conceptual model* and *conceptual design* should not be confused with the terms *concept model, concept design,* or *design concept* used by designers of physical products such as cars, landscapes, buildings, and appliances. The latter three terms refer to an early brainstorming phase, in which designers build prototypes, sketches, storyboards, or mock-ups made of cardboard, foam-core, clay, or wood, to inspire innovative, "out-of-the-box" solutions to design problems (Lombardi 2008, Buxton 2007).

The purpose of product designers' *concept designs* is to probe possible designs for the product by letting people see, touch, and otherwise experience them. The probe can explore any aspect of the design that affects users. It can explore the product's conceptual design, but most often a probe concerns the product's appearance or presentation.

For example, designers of innovative handheld appliances often create foam-core "concept models" of several alternative physical designs and have people pretend to use them. Car designers often build partially functional "concept cars" to test radically new physical designs.

Concept design is a useful technique for exploring design spaces. However, it is mainly about presentation, and as stated in Chap. 2, conflating conceptual design and presentational design can be dangerous. Therefore, we believe that presentation-focused *concept* prototypes are most useful *after* designers have developed a task-focused *conceptual* model.

3.4   Summary: Our View of Conceptual Models

3.4.1   Conceptual Models *Are*

- A method to help designers clarify, simplify, and understand what an app **is** *semantically* before designing its *presentation* (e.g., by sketching wireframes) or keystroke-level *actions* (e.g., tapping, swiping, clicking, or speaking).
- High-level object-based descriptions of an application's functionality, including the attributes, operations, and relationships of each object-type, and mappings of those concepts to user tasks. The objects are restricted to user-manipulable data.
- A basis for mental models that designers want the application's users to internalize.
- A basis for other elements of the design and development process: user interface design, usability testing, software implementation, quality testing, documentation, and training.
- A tool for communicating the design to team members, intended users, and other stakeholders.

3.4.2   Conceptual Models Are *Not*

- Not attempts to capture users' pre-existing **mental models**. Quite the opposite: a conceptual model is an **idealized, task-focused, coherent** model of the application that designers want users to internalize as their mental models.

- Not models of purely internal, implementational aspects of applications—details end-users need not be aware of—such as databases, I/O queues, network protocols, algorithms, etc.
- Not user-interface-based descriptions or user-interface prototypes.
- Not design metaphors.
- Not formal specifications to automatically generate or simulate applications. That is a goal for some who work on conceptual models, but it is not our goal.
- Not just task-based descriptions, task scenarios, use-cases, or task models. Conceptual models can include these but specify more.
- Not just information architecture and content ontologies. Conceptual models can include this but include functional as well as structural relationships between user-manipulable objects.
- Not implementation-based descriptions.
- Not *concept models*. The term *conceptual model* should not be confused with the physical product design term *concept model*. The latter refers to physical prototypes of products.

Components of a Conceptual Model

4

A conceptual model (CM) for an application should specify several things about the application: a high-level statement of the purpose and planned functionality, the major concepts and vocabulary of the application, the model's structure embodied in an Objects/ Operations analysis, and a list of unresolved conceptual model issues.[1] In this chapter, we describe each of these components in detail.

4.1 Domain, Purpose, and High-Level Functionality

A conceptual model for a software application should include a brief, very-high-level description of the application's target domain, purpose, and functionality. For example, the following is a description of the domain, purpose, and high-level functionality of a personal calendar application:

The target domain of the Personal Calendar is personal time-management. The high-level purpose is to allow people to keep track of their appointments and to-do items, possibly keeping work-related items from personal ones. Users of the calendar application can:

- Create one or more calendars for keeping track of events and to-dos,
- Schedule events, avoid or be warned about double-bookings. Delete events.
- Invite other people to events.
- View the time of events and print paper copies of events or time-periods.

---

[1] Some designers include other components, such as the effects of operations on objects, or constraints between concepts, but we consider those optional.

© The Author(s), under exclusive license to Springer Nature Switzerland AG 2024   37
J. Johnson and A. Henderson, *Conceptual Models*, Synthesis Lectures
on Human-Centered Informatics, https://doi.org/10.1007/978-3-031-50852-3_4

- Search for events (past or future).
- Create or delete to-do items.
- View to-do items.

4.2   Concepts and Vocabulary

A second component of most conceptual models is a representation of its *structure*: an enumeration of important concepts embodied by the application, and how the concepts are related. If the enumeration includes the agreed-upon user-visible names for the concepts, then it also serves as a *vocabulary*[2] for the application and its documentation (see also Chap. 10).

For example, the following is a partial list of major concepts (and vocabulary) embodied in a simple office calendar (concepts marked in **boldface**). Objects are listed in alphabetical order because this list serves as a vocabulary as well as enumerating important concepts:

- **Calendar**: This calendar. This concept is needed because the application allows users to have more than one calendar, and different users sharing the application may have separate calendars. Calendars have attributes such as: owner (a person), purpose (work, home), current focus (day, month, year). Operations could include: create, examine, change scope, print add/delete event.
- **Event**: An event recorded in the calendar. Attributes might be: name, description, date, time, duration, location, reminder, repeat (yes, no), type (meeting, birthday, vacation, personal appointment), and invitees (list of persons). Operations could be: examine, print, edit attributes.
- **To-Do Item**: Similar to events because they don't have an assigned time, but they often have a deadline. The attributes might be: name, description, deadline, priority status, and reminder. Possible operations: view, print, edit attributes.
- **Person**: A user of the application or a potential invitee to an event. Attributes: name, job-description, office, phone. Operations: contact, view details.

4.3   Structure: Objects/Operations Analysis

The most important component of a conceptual model is an Objects/Operations analysis: an enumeration, taxonomy, and organization of all concepts that the application exposes to its users. This includes the conceptual *objects* that users manipulate, *attributes* of those

---

[2] Also sometimes called terminology, nomenclature, or lexicon.

objects, the *operations* that users can perform on each of those objects, and any *relationships* between objects (Newman and Sproull 1973; Johnson et al. 1989; Card 1996).[3] Such a taxonomy is one component of the application's *information architecture* or *ontology*.

The names for objects, attributes, and operations in a conceptual model will likely evolve over the course of the project and need not reflect the terms presented to users in the final product. However, at any point in time, the terminology should be agreed upon by all team members and used consistently to avoid confusion and misunderstandings among team members and eventually among the application's users.[4] Also, operations in a conceptual model should be named from *users'* point of view (e.g., *browse* today's events), not from the *application's* point of view (e.g., *present* today's events).

Using the previous chapter's discussion of a conceptual model for a simple alarm clock, the objects, attributes, and operations would be something like this:

- **Clock**: Attributes: current time of day. Operations: view current time; set current time.
- **Alarm**: Attributes: ON/OFF, alarm time. Operations: view/switch ON/OFF, view/ set alarm time.

For our calendar example, the Objects/Operations structure could be something like what is shown in Fig. 4.1.[5] If a calendar application were built based on that conceptual structure, it would imply several things about the application:

- The only conceptual objects exposed to users would be Calendar, Event, To-Do Item, and Person.
- All types of events—meetings, birthdays, vacations, and personal appointments— would have the same attributes and operations. That might simplify learning how to use the calendar but might not make sense in some cases, e.g., vacations seldom need lists of invited people.
- To-Do Items are not Events; they are a different type of item because they have different attributes and operations.

4.3.1 Declares Concepts that the Application Will Expose

As stated above, an objects/operations analysis presents all concepts that an application exposes to its users. That means that the conceptual model should be treated as a binding

---

[3] Some design teams find it useful to also include the *parameters* of each operation.

[4] The need for consistent naming of concepts throughout the application and its documentation and training is discussed further in Chaps. 7 and 10.

[5] See Chap. 5 for several different ways to illustrate Objects/Operations analyses.

Objects	Attributes	Operations
Calendar	title, owner, current focus (day, month, year), events, to-do items	browse, print, create, find event, add/delete event, change focus
Event	name, description, date, time, duration, location, repeat (yes/no), type (meeting, birthday, vacation, personal appointment), invitees (list of persons), reminder	examine, print, edit attributes
To-Do item	name, description, deadline, priority, reminder, status	examine, print, edit attributes, mark complete
Person	name, job-description, office, phone, email	contact (email, phone, text), view/edit attributes

Fig. 4.1 Objects/Operations analysis for a simple calendar application

contract that *declares* concepts that the corresponding application exposes to its users.[6] The application should expose only objects, attributes, operations, and relationships that are in the conceptual model. Stated as design rule:

If a concept is not in the conceptual model, it must not be exposed to users.

If a design team discovers during UI design or implementation that the application must expose concepts that are not in the conceptual model, the conceptual model—i.e., the design contract—must first be revised to include the new concept, with the whole team's agreement.

In our alarm clock example, the method by which the clock triggers the alarm at the appropriate time is of no concern to users; they care only that the clock triggers the alarm at the specified time. Therefore, the alarm-triggering method is excluded from the conceptual model, and therefore from the clock's user interface.

4.3.2   Introduces New Concepts, if Needed

Computer-based products and services often provide capabilities that pre-computerized tools and methods did not provide, so concepts new to a task domain sometimes appear in the conceptual model for a new product or service. Some examples:

---

[6] Users engage the application through its user interface (UI) and documentation, so the UI and documentation are how the application's conceptual model is exposed.

- Most software-based calendars provide repeating events, in contrast to paper calendars, which require people to record recurring events by writing the same event on multiple dates.
- In a physical filing drawer, each hardcopy document can be filed in only one place, whereas electronic documents in a computerized document system can be filed in multiple folders simultaneously.
- Modern digital audio recorders allow recordings to be played back at different speeds either just like an old-fashioned tape recorder—altering the pitch of the recording—or in a way that keeps the original pitch of the recorded sounds.

However, every new concept comes at a cost, for two reasons:

- It adds an unfamiliar concept that users must learn.
- It potentially interacts with every other concept in the system. As concepts are added to a system, the complexity of the system rises not linearly, but multiplicatively.

Therefore, additional concepts should be resisted, and admitted into the conceptual model only when they meet functional requirements for the application and their extra cost can be minimized through good user interface design. Remember: less is more!

4.3.3   Shows Relationships Between Concepts

Enumerating the objects, attributes, and operations for an application allows designers to notice relationships between objects. Making these relationships explicit in the conceptual model can make the application easier for users to understand. The following object-relationships are common across application domains.

Specialization Relationship

Different objects in a task domain may have the same operations and attributes. This suggests that those objects may all be variations of one type of object, or subtypes of a common parent object. In such a case, objects in a conceptual model can often be organized in a *specialization* or *type/subtype hierarchy*, in which certain conceptual objects are *specializations* of others.[7] Alternatively, similar objects can be represented as a single object-type with an attribute indicating which of the possible subtypes they are.

Whichever approach is used, making the similarity relationship explicit in the conceptual model can help users learn more easily about all the conceptual objects. Most people can understand the idea of specialization: for example, a **savings account** is one type of

---

[7] Specialization is also a common relationship between objects in object-oriented analysis, a common early step in object-oriented programming. However, type/subtype relationships in conceptual model design only concern objects that the application exposes to users.

bank account, a **book** is one type of **product** a store might sell, a **cookbook** is one type of **book**, and a **truck** is one type of **vehicle**. Furthermore, designers can take advantage of the commonalities across similar objects: they can use the same interaction design for common operations and the same design for viewing and setting object attributes.

For example, consider a drawing application that provides functions for drawing both **rectangles** and **ellipses**. If both **rectangles** and **ellipses** have similar attributes and operations, once a user understands one, they also understand the other. As a specific example, once users learn that they can constrain **rectangles** to be squares, they should be able to immediately assume that they can similarly constrain **ellipses** to be circles. Such consistencies result in a conceptual model—and an application—that appears to have fewer distinct concepts, is simpler and more coherent, and is more easily mastered.

In the Objects/Operations analysis for a simple calendar discussed above, **meetings**, **vacations**, **birthdays**, and **personal appointments** were all modeled as **events**, with a **type** attribute indicating the specific **event** subtype, e.g., *type* = *birthday* (see Fig. 4.1, earlier in this chapter). This is a simple model that would make sense if all types of **events** have (mostly) the same attributes and operations. However, if the different types of events have extra attributes and operations that are not shared, one might model them differently, as subtype *objects* of **event** (see Fig. 4.2). They would inherit the attributes and operations of **event** but could each have their own attributes and operations.

Objects	Attributes	Operations
Calendar	title, owner, current focus (day, month, year), events, to-do items	browse, print, create, find event, add/delete event, change focus
Event (subtypes indented below)	name, description, date, time, duration, location, repeat (yes/no), reminder	examine, print, edit attributes
Meeting	host, invitees, call-in #, access code	announce, cancel
Vacation	start date, end date	
Personal appointment	private?	
Birthday	invitees (list of persons), surprise?	announce, cancel
To-Do item	name, description, deadline, priority, reminder, status	examine, print, edit attributes, mark complete
Person	name, job-description, office, phone, email	contact (email, phone, text), view/edit attributes

Fig. 4.2 Objects/Operations analysis for a simple calendar with subtypes of Event

A calendar application based on the conceptual model shown in Fig. 4.2 would differ in several ways from one based on the model shown in Fig. 4.1:

- Only Meetings and Birthdays have Invitees; Vacations and Personal Appointments do not.
- Meetings and Birthdays can be announced and cancelled. Vacations and Personal Appointments can be created and deleted but not explicitly announced or canceled.
- Vacations have a starting date and an ending date. Other types of Events are assumed to start and end on the same day.
- Personal Appointments can be made private.

Containment Relationship

In many applications, objects are also related by a second hierarchy: *containment*, in which some objects can *contain* other objects. For example:

- a **print-queue** *contains* **documents** to be printed.
- a **shopping cart** *contains* **products** a customer wants to purchase.
- a **friends list** contains **people** a user is connected to on social media.
- an **email folder** *contains* **email messages**.
- an **organization** *contains* **employees**.

If objects in a conceptual model are related by containment, the model should indicate whether the containment relationship is mutually exclusive or not, that is, whether an object can be in multiple containers at once or only in one container at a time (see Chap. 7).

Part-of Relationship

This is like containment, but different. For example, a photo album contains photos, but they are not part of the album itself. A bus *contains* passengers, but they aren't parts of the bus. The wheels, motor, and windows, on the other hand, *are* parts of the bus.

Sometimes, it is hard to see the difference, e.g., a chapter is both contained in a book and part of it, and a paragraph is both contained in a document and part of it. The part-of relationship is similar enough to containment that some designers model them as the same relationship. This is discussed further in Chap. 7.

Other Common Object Relationships

A conceptual model for an application can also specify other relationships between objects if they are important to the tasks the application is intended to support. Examples of relationships sometimes made explicit in conceptual models:

- *Task/sub-task*: One task may be a part of a larger task, e.g., opening a bank account consists of several steps, each of which can be regarded as a task on its own. This relationship is between operations, rather than between objects. This relationship usually is discovered during task analysis (see below).
- *Source/result*: One object may be a source, e.g., data used by an operation, while another is the result, e.g., the data changed or produced by the operation. For example, in a calendar application, the operation of inviting people to a previously scheduled meeting would have the meeting and several person-objects as *sources* and yield an updated meeting as a *result*. In a photo-editing application, a "raw" digital image would be a *source* to an editing operation, and an adjusted and cropped final photograph would be the *result*.
- *Importance*: Concepts in a conceptual model—objects, attributes, and operations—can differ in *importance*. Some concepts are encountered by users more frequently than others. For example, closing a checking account is an infrequent operation compared to, say, entering a transaction into an account. The relative importance can be used to focus the user interface design: it is more important to make frequent operations easy, even at the expense of less-frequent ones. In contrast, less-frequent operations can require more keystrokes, but should be easy to remember from one time to the next (Johnson 2020).

4.4   Conceptual Scenarios

Many designers include one or more conceptual scenarios as a component of conceptual models. Such scenarios describe people using the application. They are written in terms of the objects, attributes, and operations of the conceptual model, *not* in terms of a user interface. Conceptual scenarios are a good way to describe a conceptual model to non-technical stakeholders, and even to intended users of the planned application.

Conceptual scenarios are also useful to designers because they describe *task-to-tool mappings*: how people combine operations provided by an application to perform tasks in the application's target domain. As discussed in Chap. 2, a person with a task devises a plan consisting of operations provided by the application—and possibly other activities that they do outside of the application (including using other applications)—to accomplish their task.

Conceptual scenarios help designers evaluate a conceptual model by demonstrating how easily various tasks can be performed. A conceptual model that fits its domain well allows the application's users to accomplish simple or common tasks with simple plans: concepts in the task-domain map directly to concepts in the conceptual model, so the necessary sequence of operations is clear and short. Plans for less-common tasks may be more complex, requiring more forethought and/or more operations. If scenarios show a need for people to form complex plans to complete simple or common tasks, the conceptual model is deficient and should be redesigned.

For example, most home and small-business photocopiers are designed so that the task-to-tool mapping (plan) for making one copy of a one-page original is simple: put the page in the original tray, push start, take the copy out of the copy tray, and take your page out of the original tray (UI hint: remind people not to forget the last step). Doing a complex copying job using a simple copier requires a more complex plan, involving activities done apart from the copier, such as collating or stapling pages by hand.

In contrast, in the reprographics shop of a large business, simple copying tasks are rare. Doing a simple copying job—e.g., one copy of one page—on a big commercial photocopier may require a complex plan: an operator might have to adjust many settings to set up a "simple" copying job. That is OK in that environment. However, the more-common complex tasks—e.g., 100 copies of a multi-page, two-sided, color document—should require little planning or set-up by the trained operators.

The following is a conceptual scenario for an address book application. Notice that it includes no presentation or interaction details: no pages, screens, or windows; no menus, buttons, sliders, or tabs; no typing, scrolling, tapping, or speaking. The same scenario could apply to a desktop app, a smartphone app, a Web-app, or a voice-controlled app for a "smart" speaker or for sight-impaired people.

> Sally needs to update addresses in her address book. She opens the app, finds the entry for her friends who moved across town, opens it for editing, edits it, reviews it for accuracy, and saves it. Another family she knows got rid of their land-line, so she edits their entry to delete that phone number. Similarly, she edits three entries for friends who recently moved, merges entries for two friends who got married and now live together, and separates an entry for two other friends who split up. While doing all this, Sally notices entries that she no longer needs in her address book, and deletes them. After making sure her updated address book will be synchronized across her various devices - mobile phone, tablet computer, laptop computer, and desktop computer - she quits the app.

Notice also that conceptual scenarios can be separated into brief use-cases, which can be assigned to developers to implement.

4.5   Conceptual Model Issues

A final important component of a conceptual model is a list of resolved and still-open conceptual design issues, i.e., known problems in the conceptual model. If an issue has been resolved, it should be so marked (or moved to a separate list of resolved issues), and the resolution should be recorded with the issue, so newcomers to the design project can see why the design is as it is.

The following are examples of Resolved and Open issues from our hypothetical simple calendar application:

4.6   Resolved Issues

1. The types of Event don't all share the same attributes and operations. *Resolution: Model them as subtypes of Event, so they inherit the attributes and operations of Event but have additional attributes and operations.*
2. Do Vacation, Personal Appointment, and Birthday (subtypes of Event) have operations other than those they inherit from Event? *Resolution: No.*

4.7   Open Issues

1. Should Reminder be an object, rather than just an attribute of Event and To-Do Item? Is being considered because one could view Reminders as having attributes of their own. If so, Event and To-Do would still have reminder attributes, with a Reminder object as the value of the attribute.
2. Should the Contact operation of Person be split into separate operations?

Representing Conceptual Models

5

5.1   Introduction

We hope that by now we have convinced you that designing a conceptual model is a useful step for clarifying, understanding, organizing, and simplifying the structure, content, and function of the software you are designing. Conceptual models must also convey that understanding to all team members and stakeholders, including those who lack technical training. Thus, an easily understood and informative representation of a conceptual model is an important deliverable.

As Chap. 4 explains, conceptual models have several components: a statement of purpose and high-level functionality, a list of major concepts and terminology, the model's structure based on an Objects/Operations analysis, scenarios, and resolved and unresolved issues. All components except the Objects/Operations structure are easily represented textually, so this chapter focuses on ways to represent the Objects/Operations structure component.

Objects/Operations structures can be represented in the form of an outline consisting of object-types and subtypes, each with operations and attributes. They can be represented in tables or spreadsheets, with objects in rows and actions and attributes in columns.

Alternatively, Objects/Operations structures can be encoded in a computer-interpretable modeling system, such as diagrams in the Unified Modeling Language (UML). Once a model is complete, some such systems can generate code implementing a skeletal prototype of the application. Programmers then add code to the prototype to implement detailed behavior.

A conceptual model's Objects/Operations structure can also be represented by other types of diagrams, such as object-relationship diagrams (commonly known as entity-relationship or ER diagrams) and concept maps.

© The Author(s), under exclusive license to Springer Nature Switzerland AG 2024   47
J. Johnson and A. Henderson, *Conceptual Models*, Synthesis Lectures
on Human-Centered Informatics, https://doi.org/10.1007/978-3-031-50852-3_5

Finally, a conceptual model's objects, attributes, and operations can be represented in a conceptual scenario format. The scenario connects short descriptions of someone using the application (called use-cases), covering all that users can do with the application. It is written at a purely conceptual level, using terms from the conceptual model. The scenario contains no user interface or implementation details.

In our experience, when a conceptual model is shown to team members to get their feedback, it is usually best to represent the Objects/Operations structure as an outline, a spreadsheet, or an entity-relationship diagram, because those presentations are most easily understood by everyone on the design-development team.

The remainder of this chapter provides illustrations of the above-described common ways to represent the Objects/Operations structure of conceptual models.

5.2   Outline

To represent Objects/Operations structures in outline format, it is necessary to follow some typographical conventions. Here are the conventions we recommend:

- Objects names are **bold**. Attribute names and values are italicized. Attribute values that are objects are ***bold italics***.
- Each object has "slots" for subtypes, attributes, and operations. In some objects, some of these slots are empty, that is, some objects have no subtypes, no attributes, or no operations.
- Objects that are contained in other objects indicate their container object type in a Container slot.
- Objects that have parent-types inherit attributes and operations from their parent unless they redefine the attribute or operation.
- Some operations have sub-actions, or variations. They are indented below the main operation.

Figure 5.1 shows an example of an Objects/Operations structure for a simple calendar application like the one discussed in Chaps. 4 and 5.

5.3   Table or Spreadsheet

Figure 5.2 presents the same Objects/Operations structure as Fig. 5.1, this time using a table.

- **Calendar** (one per user)
  Parent types: none
  Subtypes: none
  Attributes: *Title: <text>; Owner: <**Person**>; Current focus: <day, week, or month>, Events, To-Do Items*
  Operations: examine, print, add **Event**, delete **Event**, change focus

- **Event**
  Parent types: none
  Subtypes: **Meeting, Vacation, Personal appointment, Birthday**
  Attributes: *Name: <text>; Description: <text>; Date: <day, month, year>; Time; Duration; Location; Repeat: <yes/**no**>, Reminder*
  Operations: examine, print, edit attributes

  - **Meeting**
    Parent type: **Event**
    Subtypes: none
    Attributes (beyond inherited ones): Host: **Person**; Invitees: <list of **Persons**>; call-in #; access code
    Operations (beyond inherited ones): *announce, cancel*

  - **Vacation**
    Parent type: **Event**
    Subtypes: none
    Attributes (beyond inherited ones): Start date: *<day, month, year>*; End date: *<day, month, year>*
    Operations (beyond inherited ones): *none*

  - **Personal Appointment**
    Parent type: **Event**
    Subtypes: none
    Attributes (beyond inherited ones): Private: <**yes**, no>
    Operations (beyond inherited ones): *none*

  - **Birthday**
    Parent type: **Event**
    Subtypes: none
    Attributes (beyond inherited ones): Invitees: <list of **Persons**>; Surprise: <yes, **no**>
    Operations (beyond inherited ones): *announce, cancel*

- **To-Do Item**
  Parent type: none
  Subtypes: none
  Attributes: *Name, description, deadline, priority, reminder, status*
  Operations: *examine, print, edit attributes, mark complete*

- **Person**
  Parent type: none
  Subtypes: none
  Attributes: *Name, Job-description, Office, Phone, Email*
  Operations: *Send email, view/edit attributes*

Fig. 5.1 Outline representation of a conceptual model for a calendar application

5.4 Unified Modeling Language (UML) Diagram

The Unified Modeling Language was developed mainly to allow software programmers to model class hierarchies for object-oriented software, but it can also be used to represent Objects/Operations structures in conceptual models (see Fig. 5.3).

UML editors are available that allow designers to create conceptual models in the form of UML diagrams. Some such editing tools can then translate the UML diagrams into

Objects	Attributes	Operations
Calendar	title, owner, current focus (day, month, year), events, to-do items	browse, print, create, find event, add/delete event, change focus
Event (subtypes indented below)	name, description, date, time, duration, location, repeat (yes/no), reminder	examine, print, edit attributes
Meeting	host, invitees, call-in #, access code	announce, cancel
Vacation	start date, end date	
Personal appointment	private?	
Birthday	invitees (list of persons), surprise?	announce, cancel
To-Do item	name, description, deadline, priority, reminder, status	examine, print, edit attributes, mark complete
Person	name, job-description, office, phone, email	contact (email, phone, text), view/edit attributes

Fig. 5.2   Tabular representation of a conceptual model for a calendar application

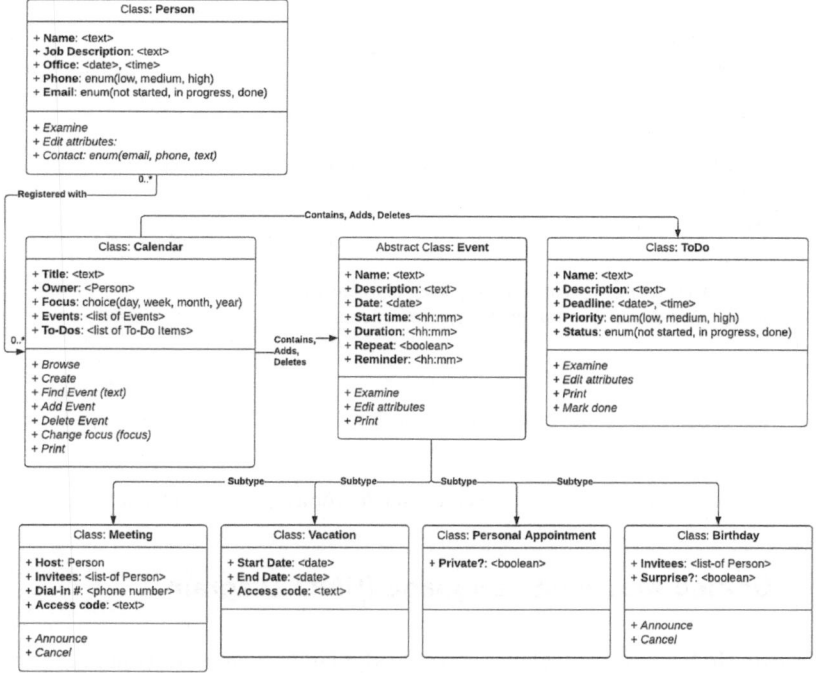

Fig. 5.3   UML class diagram of a conceptual model for a calendar application

application source code, so representing a model in UML can facilitate the development of code that implements the conceptual model. However, representing a conceptual design in UML diagrams may *hinder* understanding of the model by non-programmer members of a development team. It could also increase the likelihood that programmers on the team will want to include implementation concepts in the conceptual model so they can be automatically generated, which could damage the user-task focus of the conceptual model.

5.5 Entity-Relationship (ER) Diagram or Concept Map

When representing *relationships between objects* is as important or more important than the objects themselves, designers often represent Objects/Operations structures using Entity-Relationship (ER) diagrams in which the conceptual objects are the entities. Concept maps, invented to help people visualize connections between collections of ideas and concepts, are similar to ER diagrams and can also be used to represent relationships between objects and effects of operations in conceptual models. Figure 5.4 shows an ER diagram (or a concept map) for the Objects/Operations structure of a simple office calendar.

5.6 Conceptual Scenario

As discussed in Chap. 4, a good way to present a conceptual model to non-technical team members—and even to users—is to show scenarios of people using the application. Conceptual scenarios mainly show how tasks in the application's domain can be translated or mapped to operations in the conceptual model, but they also help illustrate a conceptual model's Objects/Operations structure. In many cases, conceptual scenarios have already been provided as a component of the model.

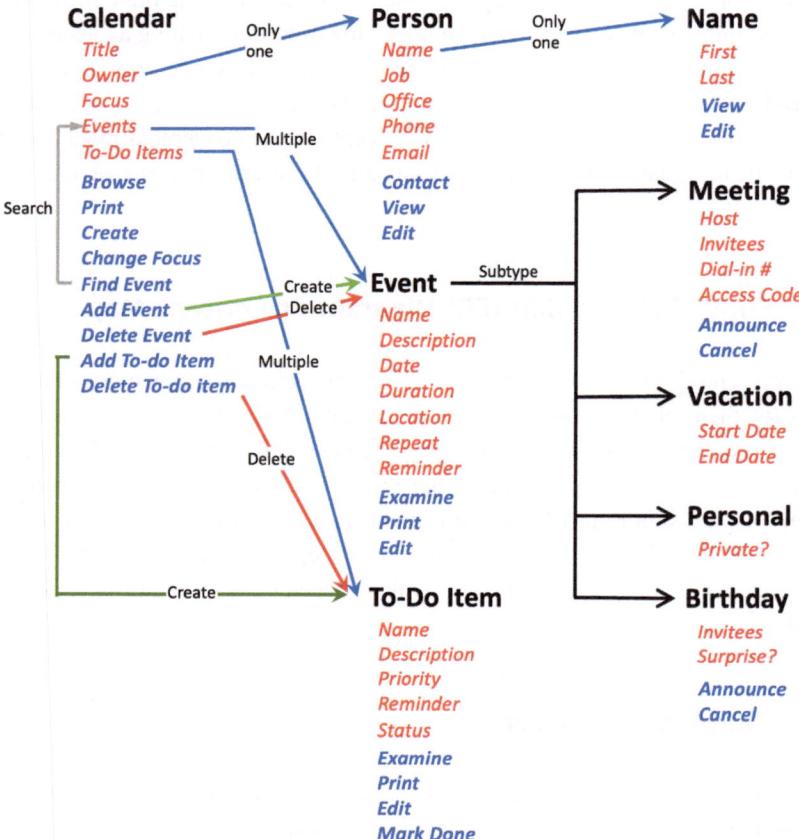

Fig. 5.4 Calendar conceptual model represented by ER diagram or Concept Map. Objects are **black and large**, attributes are *red and italicized*, and operations are *blue, italicized, and bold*

The following is a scenario for the Objects/Operations component of the simple calendar conceptual model discussed above. Objects are highlighted in **black bold**, attributes are *italicized in red*, and operations are in *blue bold italics*.

Sam wants to review his upcoming **events** and **appointments**. He opens his **Calendar** application and browses this week's **events**. He stops browsing to check the *location* of a medical **appointment** he has this Monday. He doesn't want to forget that **appointment**, so he *sets* a *reminder* to signal him one hour ahead of the **appointment**. Sam wants to see when his previous medical **appointment** was, so he *searches* the **calendar** for his doctor's name… and finds that his last doctor appointment had a *date* of two months ago. During that visit, Sam's doctor advised him to start taking low-dose aspirin, but he hasn't yet done it, so he *adds* a **To-Do item** to buy 81mg aspirin tablets. While *adding* that, he notices another **To-Do item** that he has already done, so he *marks* its *status* complete. Returning to this week's **events**, Sam changes the **calendar's** *focus* from weekly to monthly so he can quickly skip ahead to December to find out when Hannukah is this year. He *adds* a dinner **appointment** on the second day of Hannukah and *invites* three **people**. Then Sam closes the **calendar**.

Notice that this story uses conceptual-level terms only. It says nothing about the calendar's presentation, operation, or implementation. No clicking, swiping, or scrolling; no menus, buttons, links, pages, or icons. The story uses verbs like "finds" and "notices" instead of "sees". The conceptual model even uses "focus" instead of "view" for the setting that controls whether the calendar is presenting days, months, or years. The term "view" refers to a visual presentation, which is a UI detail. The conceptual model must work even if the application is accessed via a personal digital assistant with no screen, such as Alexa or Cortana, so "focus" is more appropriate than "view".

Two Complete Examples

6

To make the ideas presented in the previous chapters more concrete, this chapter presents two examples of conceptual models. Previous chapters provide examples of *parts* of conceptual models, most of which are for hypothetical simple applications. In contrast, the examples presented here are real and fairly complete. They are similar to conceptual model documents we prepare for clients. Those documents are structured as follows:

- Domain and overall purpose
- High-level functionality
- Major concepts and vocabulary
- Objects/operations structure
- Task hierarchy enumeration (optional; Example 1 only)
- Conceptual Scenarios
- Resolved conceptual design issues
- Unresolved conceptual design issues.

6.1 Example 1: Server Management Console

Our first example is a conceptual model for a server management console application.

© The Author(s), under exclusive license to Springer Nature Switzerland AG 2024
J. Johnson and A. Henderson, *Conceptual Models*, Synthesis Lectures
on Human-Centered Informatics, https://doi.org/10.1007/978-3-031-50852-3_6

6.1.1   Purpose

Domain: computer server monitoring, control, and management. Used by system operators and administrators to monitor and manage the operation of banks of networked computer servers.

6.1.2   High-Level Functionality

The Management Console allows server system administrators and operators to:

- initially install, set up, and configure servers as desired,
- operate servers and monitor server operation, throughput, and health,
- reconfigure servers during operation, e.g., to reallocate resources as needed,
- take servers offline for maintenance and upgrades,
- diagnose problems as they arise, and either fix them or obtain information to provide to people who can fix the problems.

6.1.3   Major Concepts and Vocabulary

The following is a partial list of major concepts (and vocabulary) embodied in a server management console (concepts marked in **boldface**):

Alarm: A notification that an event has occurred that triggered an Alarm Rule. Alarms have attributes such as: **time-of-occurrence**, **type**, **severity**, and **status**. Alarms are persistent: they remain in effect until dismissed or the condition that triggered them is no longer true.

Alarm Rule: A rule specifying criteria for triggering an Alarm. Criteria can include **monitored variables** that exceed a threshold value, events that exceed a threshold frequency, etc.

Enterprise: Defines a boundary beyond which resources are not shared. An enterprise often corresponds to a company, as when a server-farm hosts servers for many different customer companies and doesn't want any resources or information to cross company boundaries for legal or accounting purposes. Within a company, Enterprises may also be used to separate resources for company divisions.

Environment Probe: A sensor placed in a Server that measures—and thereby permits monitoring of—an important **monitored variable**, e.g., temperature, humidity, voltage.

Event: Anything that is detected by the Service Processor. Events have a **Severity** attribute that indicates the event's importance: Normal, Warning, Problem. All events are recorded in a log.

Field Replaceable Unit (FRU): Hardware component that can be removed from the server at a customer site and replaced (if necessary). Not all parts of a server are FRUs; some parts can be replaced only at the factory.

Non-FRU-Part: Components that aren't Field Replaceable Units (FRUs).

Log: A facility of the Management Console that records events and lists them for review by users. The Console will maintain several logs for different purposes.

Log Entry: A single item in a log, documenting one **Event**. Each log entry has a textual **description**, a **severity**, a **time**, a **source**, and other pertinent data.

User: A person who uses the Management Console (i.e., any of its functions). User attributes include: **name**, **authentication information** (login, password), **manager**, and current **role**(s). Most users will have a single fixed Role, e.g., web-server operators, system administrators, hardware technicians, or managers, but some may have multiple Roles or may change Roles.

Role: A specific administrative job-function for a **User**, with specific permissions attached to it. In one Role, a given user will be allowed to perform certain functions but not others. In another Role, the same person might have different permissions.

Server: A computer that provides one or more services such as file, email, web-hosting, database, proxy, print, virtual machine, name, and application.

Objects/Operations Structure
For this conceptual model, we use a table to present the Objects/Operations structure (see Fig. 6.1).

6.1.4   Conceptual-Level Scenarios

1. A severity 5 alarm occurs on 2024/07/18 at 03:06. The on-duty sysadmin notices it and checks to see what rule triggered the alarm. The sysadmin checks the log for log entries indicating the cause and possibly relevant prior events. The cause appears to be a crash in server #4076, caused by a faulty cable. The faulty cable is field-replaceable, so the sysadmin goes to the server, shuts it down, replaces the cable, clears the alarm, and restarts the server. The alarm remains cleared, indicating that the repair was successful.
2. The manager of the server bank logs into the console. Her management role lets her access an overview of the entire server bank that her sysadmin employees cannot. She notices that most of the servers are functioning, but logs show that a few are almost

Objects	Attributes	Operations
Enterprise	Name, # of servers, authorized users, manager	View/edit authorized users, View/edit attributes
Server	ID, model, Services: (list of service types),	install, remove, upgrade, activate/deactivate
Field Replaceable Unit (FRU)	asset #, server (0 or #)	add, remove, view attributes
Non-FRU Part	asset #, server (0 or #)	add, remove, view attributes
Event	severity: (1-9), server ID,	view
Environmental Probe	location, type, status	add, remove, view attributes
Alarm	time, date, severity, status	mute/unmute, view attributes
Alarm Rule	name, trigger, effect, date added, triggered count	view, create, edit, delete
Log	type, log entries	view, search, archive, clear
Log Entry	time, date, content	post, view, edit, delete
User	name, ID, contact info, authorized roles, experience	view/edit contact info
Role	type, qualifications	(mgr only) view/edit, delete

Fig. 6.1 Objects/operations analysis for a server management console

overloaded. She also notices that one of the five sysadmins has not yet logged in. She sends a text to that sysadmin. While awaiting a reply, she initiates an environmental probe to try to discover why some servers are nearly pegged.

6.1.5   Resolved Conceptual Design Issues

1. What is the scope of a log? Are logs global to the entire Management Console, or are they specific to Enterprises? _Resolution: In principle, logs are scoped by Enterprise, but since version 1 has only one Enterprise, version 1 can be regarded as having only one log, which is global._
2. How is the "source" of an Event identified? _Resolution: Component ID, location._

6.1.6   Open Conceptual Design Issues

1. The log will record huge numbers of events, so it will be hard for users to find and keep track of important problem-events. One solution is to have separate logs for normal

versus problem-events, and the ability for users to delete problem-events once handled. Another possible solution is a filtering mechanism for the log, allowing users to view only small subsets of interesting events.

2. The model has a concept of an Enterprise: an enclosed, protected environment in which services are run. One issue is whether we also need the concept of a Hardware Domain: a division of a physical server's hardware resources, upon which Enterprises run. If users need to know which physical machines their Domains have been placed on, the concept of Hardware Domains may be needed.

6.2   Example 2: Web-Search Service for Professionals

The second example is a conceptual design for a high-end Web-Search service intended mainly for professional information workers. It is adapted from an actual design done several years ago, so the name of the service has been changed and details have been changed or omitted for the sake of brevity, clarity, and intellectual property protection.

6.2.1   Domain and Purpose

Domain: Professional advanced search. The purpose of the **DocLivingstone** service is to allow people, mainly professional information workers (e.g., librarians, travel agents, patent attorneys) to search for information on the Web in a more systematic, sustained, comprehensive, and flexible way than most search services support. Search in DocLivingstone is systematic, sustained, and comprehensive because search projects span multiple individual searches and even multiple usage sessions, retaining the information collected until it is no longer needed. Search in DocLivingstone is flexible because DocLivingstone accepts as search input not only keywords, but also documents and web URLS.

6.2.2   High-Level Functionality

The DocLivingstone service will provide the following functionality:

- DocLivingstone users can create Search projects (called **Explorations**), which persist across individual searches and use-sessions. Explorations can be started, worked on, set aside, returned to, and resumed later. A user may have several Explorations at any point in time, and may shift from working on one to another at will. Explorations hold information that the user wishes to retain. They can be deleted when no longer needed.

- As input to a search, users can provide text (search terms or prose), webpage URLs, textual documents, or a combination of these.
- Documents returned by searches (formerly called Search Result Items) can be organized in different ways, e.g., as an ordered list (table) of individual found documents (like results in most search engines), or by common topics extracted from the documents.
- Users can refine searches by indicating that certain found documents are especially relevant. DocLivingstone uses this to revise the search, thereby modifying the search results.
- Specialized **Explorations** (e.g., business travel, patent search) are provided for certain common use-cases. These guide users through the steps of conducting a search, and provide results-displays and analyses customized for the type of **exploration**.

6.2.3   Major Concepts and Vocabulary

This section defines terms for the service. The following terms should be used consistently in the service and all documentation about it.

DocLivingstone: The service itself. Needed because users need a way to refer to it.

Exploration: A container for a user's search-work in pursuit of an information goal and the results it generates. Contents of **explorations** include Search input, topics, results lists, found documents (aka results items), notes, and analysis results. An **exploration** is a way of organizing a search, refinements, and the resulting data according to the high-level goal a user is trying to achieve (similar to *projects* in other applications). An **exploration** can contain multiple Search Results Lists (see below). Each Search Results List documents the input on which it is based.

Exploration List: A list kept by the service of a user's current **explorations**.

Search Input: Information the user supplies to an **exploration** to indicate what s/he is looking for. Input to a search consists of text, which can be supplied in a variety of different ways: typing search terms, typing full sentences or paragraphs, specifying documents, and indicating web-pages. Search Input may consist of one or more Search Input Items.

Search Input Item: A single piece of input to a search, provided in one of the several possible ways: typed text, document, URL.

Search Results List: A list of **documents** found by a search, rank-ordered by match score, from highest match to lowest.

Document: Formerly called "Search Result Item". An item found by a search. **Documents** can be any of several media types: patent, patent application, advertisement, magazine, journal, newspaper, book, website, blog. The subject of a Document is separate from its media type.

Media Type: The medium of a **document** by a search. Media types include: newspapers, magazines, advertisements, journals, books, websites, blogs (a special type of website), patents, and others. A **document's** media type determines the attributes it is assumed to have, e.g., blogs have contributors, dates, etc.; patents have classifications, assignees, etc.

Content Category: Pre-defined categories into which all textual content is categorized. The pre-defined categories are: technology, business, entertainment, sports, science, and medicine.

Relevance Feedback: Users can provide the service with feedback on the actual relevance of a given **document** returned by a search. This information is used by the service to revise the search criteria to give higher weightings to the attributes in found items that the user marks as highly relevant, and then initiate a new search.

Topic: A phrase derived from common content in a list of Documents returned by a search, which serves as a title for that content.

6.2.4   Objects/Operations Structure

For this conceptual model, we chose to represent the Objects/Operations structure in Outline format. For the typographical conventions used in this style of presentation, please see Chap. 5.

- **Exploration**

  - Subtypes
    none

  - Attributes
    Name: text
    Description: text
    Create date: date & time
    Modify date: date & time
    Results: list of **Documents**
    Alert:
        Set: ON/OFF
        Checking Frequency: enum {hourly, daily, weekly, monthly}
        Alert Method: enum {markUserTask, email}
    Notes: text
    Operations: list of past operations in this Exploration, latest first

  - Operations
    Search for Documents matching specified Search Input
    Review Search Results
    Refine Search
    Undo last operation
    Redo last undone operation
    Copy
    Delete
    Set Alert (to recheck periodically and notify user when something new is available)
        Turn ON/OFF
        Set Checking Frequency (freq)
        Set Alert method (method)
    View Alert
    Save (only for **Explorations** *that have not yet been saved)*

- **Exploration List** (one per user)

  - Attributes
    Username: text <valid username>
  - Operations
    Clear
    Expand *Exploration*
    Contract *Exploration*

- **Search Input**

  - Attributes
    Items: List of Search Input Items
  - Operations
    Submit
    View

- **Search Input Item**

  - Subtypes
    typed text, file, web page
  - Attributes
    Content: text
  - Operations
    View

- **Typed Text Search Input Item** (subtype of Search Input Item).

  - Attributes
    TBD
  - Operations
    View: display of text

- **Filed Search Input Item** (subtype of Search Input Item)

  - Attributes
    Filename: text
  - Operations
    View: display of filename

- **Web Page Search Input Item** (subtype of Search Input Item)

  - Attributes
    URL: text
  - Operations
    View: display of URL

- **Search Result List**

  - Attributes
    Input Data: Search Input
    Description: text string
    Items: list of **Documents**
  - Display controls:
    Category shown: enum {<subtypes of Documents>}
    Number of items shown: integer
    Order by: enum {relevance, popularity, <other?>}
  - Operations
    View/Edit Attributes

- **Document (aka Search Result Item)**

  - Subtypes
    patent, advt, magazine, journal, newspaper, book, website, blog
  - Attributes
    Name: text
    Datafields: [pertinent data differs by subtype]
    Content Category: enum {technology, business, entertainment,
    sports, science, medical}
    URL: url
    Date: date
  - Operations
    Give relevance feedback
    Make input item
    View Attributes

- **Patent Document (subtype of Document)**

  <details omitted>
- **Advt Document** (subtype of Document)

  <details omitted>
- **Magazine Document** (subtype of Document)

  <details omitted>
- **Journal Document** (subtype of Document)

  <details omitted>

- **Newspaper Document** (subtype of Document)

  <details omitted>
- **Book Document** (subtype of Document)

  <details omitted>
- **Website Document** (subtype of Document)

  <details omitted>
- **Blog Document** (subtype of Document)

  <details omitted>

- **Topic**

  - Attributes
    Label: text (not user settable)
    Contents: list of **Documents**
  - Operations
    View Content documents.

6.2.5   Task Hierarchy Enumeration (from Task Analysis)

This section lists the tasks that the **DocLivingstone** service is meant to support. Strictly speaking, this is **not** part of the conceptual model. It is the results of the task analysis from which the conceptual model was generated. We include it here because we sometimes include the task analysis results when presenting conceptual models to clients.

These tasks are common to most or all of the top-level consumer searches described in the Objects/Operations structure (above). Major tasks are broken down into sub-tasks, some of which are task-steps, while other sub-tasks are alternatives.

- Provide input to search
  - Type input keywords or description
  - Provide input file
  - Provide input URL
- Start search
- Review search results
  - Navigate through Themes and Documents resulting from a search
  - Open a resulting Theme folder to view the Documents in it
  - Open a specific Document to view its content

- Act on resulting Documents
  - Save Document in Exploration's Saved Documents list
  - Mark Document as relevant to refine search results
  - Mark Document as not relevant to refine search results
- View Exploration's Saved Documents list
  - Delete Document from Saved list
- Alter Exploration Details
  - View Exploration Details
  - Change Exploration Type
  - Edit Exploration Input
- Start a new Exploration
- Switch to different Exploration
- Maintain Explorations
  - save Exploration
  - name/rename Exploration
  - provide description
  - delete Exploration(s)
- Set Alert on Exploration
- View Alert Notification.

6.2.6   Conceptual-Level Scenarios

1. The patent attorney has spent two weeks searching for possible prior art for a client's patent. When he started the search, he created an Exploration, which the DocLivingston app automatically added to the attorney's Explorations List. Today he logs into the app, and it shows him his Explorations in progress. He opens the prior-art Exploration and prepares to resume his search. Today he has a research paper on a topic similar to the patent, so he feeds the document to the app as Search Input and directs the app to look for documents similar to that one. A few seconds later, DocLivingston returns a Search Results List containing titles of eight documents that either reference the document he input, or discuss topics related to it. Two are patents, two are journal papers, one is a web page, one is a book, and two are newspaper articles. He begins browsing through the documents.
2. The next day, the lawyer resumes searching for prior art for the patent. He starts by deleting a few past searches from the Exploration, as they were not useful. This time he reads aloud a series of search terms into the search input. He sets the app to return only patents and journal articles, then starts the search.

6.2.7   Resolved Conceptual Design Issues

1. Users can provide web URLs as input to a search. Does a URL specify a web *page*—
   i.e., only the specific page having that URL—or an entire web*site*? If a URL means a
   web*site*, then how does the service know where the site ends? How many links away
   from the given URL does it scan? *Resolution: At least in the first release and perhaps
   forever, a URL provided as input to a search specifies only a single page. The search
   input mechanism does not follow links outward from the specified page.*
2. Are patents and patent applications the same concept or different concepts? *Resolution:
   Patents and patent applications will be treated as the same concept. The only difference
   between them is that patent applications have null values for certain attributes, e.g.,
   Grant Date.*
3. Within an Exploration, users can refine a search in several different ways. Each
   refinement produces a new set of results. Does the service keep the results of every
   refinement, or does it keep only the results of the latest refinement? Can users get back
   to the result of any Search they have done in a given Exploration? *Resolution: Only
   the results of the last search are kept in a given Exploration. Users refine the Search
   by indicating the relevance of items or by turning result-documents into Input Items.
   However, actions taken in an Exploration are saved, so in principle, users can back out
   of them one at a time, using something akin to UnDo.*

6.2.8   Open Conceptual Design Issues

1. If a user chooses not to organize different search projects into separate Explorations
   (i.e., ignores the concept of Explorations, treating DocLivingstone as a simple search
   tool), everything the user does after initiating a Search goes into a default Exploration.
   This includes providing relevance feedback on items, performing analyses, and other
   actions. However, if the user provides new Search input, the current design starts a new
   Exploration. It is unclear whether—or under what conditions—the first Exploration
   will be saved or overwritten by the new one. One possibility is that Explorations are
   not saved unless a user either (a) explicitly saves them, or (b) implicitly indicates
   that the Exploration is save-worthy by, e.g., renaming it something other than the
   default name. Also, a default Exploration could be saved across sessions, or it could
   be forgotten. If it is no trouble to save it automatically, it would be better to do that
   than to require users to explicitly save their work.
2. In Search Results, if a user marks a resulting document as a good result, the service
   uses that to adjust the weights of its selection and ranking criteria, so the Search Results
   list may change. However, that does not alter the search input for the Exploration.
   Thus, a given Exploration has *state* that is not reflected in its Search Input. Is that
   state visible anywhere to users? If not, it may seem mysterious that two people can
   provide the same Search Input, yet have different Search Results.

Essential Conceptual Modeling

7

Modeling software applications comprehensively and coherently is difficult. Modeling them comprehensively, coherently, and *simply* is even harder. Modeling them comprehensively, coherently, simply, and in a way that makes sense to people who want to do things with the application is *very* difficult. This chapter provides advice on how to create conceptual models that meet all of these goals.

Chapter 4 described the components of conceptual models. This chapter begins with instructions on how to perform basic objects/operations analysis, including pitfalls to avoid. Following the basics, this chapter discusses common conceptual design issues that often cannot be resolved early in design, but *must* be resolved before a conceptual model—and the application embodying it—will make sense.

Chapter 8 covers additional issues that commonly arise during conceptual design, but that are *optional*: designers can ignore them and still have a coherent conceptual model and application.

7.1 Basic Objects/Operations Analysis

Objects/Operations analysis is a bit like solving crossword or jigsaw puzzles: all necessary parts must be included, parts must fit together, and there can be no conflicts or duplications.

Designers new to conceptual models often have trouble with objects/operations analysis. They have trouble deciding (a) what the main user-manipulable objects in an application are, (b) whether a given concept should be modeled as an object, operation, or attribute, (c) what the attributes of each object are, and (d) which object each operation

J. Johnson and A. Henderson, *Conceptual Models*, Synthesis Lectures
on Human-Centered Informatics, https://doi.org/10.1007/978-3-031-50852-3_7

acts upon. For example, in a banking application, is a **deposit** into a bank account an object or an action? (Hint: it does **not** depend on whether you say, "The user deposits $10" or "The user makes a $10 deposit".) In a photo management application, is *deleting a photo* an operation on the **photo** object or on the **album** that the photo is in? Below are tips to help designers overcome some of these hurdles and perform objects/operations analysis successfully.

7.1.1 First Step: Identify the Objects, Attributes, Attribute Values, and Operations

An initial breakdown of the objects, operations, and attributes in a conceptual model can often be obtained from transcripts of worker interviews and work-observation sessions. When people tell researchers how they work, or when they converse with others while working, they tend to use certain parts of speech for each type of concept. Nouns usually refer to *objects* (e.g., bank account, meeting, photo, post) or *attributes* (e.g., name, description, date, interest rate). Adjectives and adverbs often refer to *attribute values* (e.g., red, small, bright, weekly). Verbs usually refer to *operations* (e.g., open, edit, delete, print). Thus, transcripts from user research can provide a first draft of an objects/operations analysis.

The first draft often must be adjusted, because:

- *objects* are sometimes expressed as verbs, e.g., "I tell friends where I am by *texting* them."
- *operations* are sometimes expressed as nouns, e.g., "… then I send out *invites* for the meeting".
- *attribute values* are often neither adjectives nor adverbs, e.g., Season: *Spring*, Name: *Sergei*, Severity: *3*, Departure Time: *14:30*.

Also, designers must decide which nouns in a user-interview transcript are *objects* and which are *attributes (e.g.,* Season*)* or *attribute values* (e.g., Spring). Designers new to objects/operations analysis sometimes confuse *objects* or *attributes* with *attribute values*, e.g., Fall, Spring, Summer, and Winter aren't objects or attributes; they are *values* of a Season *attribute*.

Nonetheless, capturing the nouns, verbs, and attributes from user research transcripts is a useful first step. The following sections provide guidelines to help refine a conceptual model and avoid common modeling mistakes.

7.1.2 Objects Are UI-Independent User-Manipulable Content

Conceptual objects are user-manipulable data in the application. Parts of the user interface—e.g., screens of an application—are not conceptual objects. For example, in a smartphone banking app, a user's **bank account** is a valid object, but the app's My Account page is not; it is just a user interface for viewing and manipulating the user's **bank account**. Similarly, website landing pages (e.g., Favorites or Contact Us) are not objects. For example, a conceptual model of a web-shopping site may have **shopping cart** as an object, but the shopping cart *page* is only a user interface for accessing the user's **shopping cart** object.

7.1.3 Assign Operations and Attributes to Objects

Any objects in the model are there because users can *do* something with them. A conceptual model should not include objects that no user can act upon in some way. If an object in a conceptual model has no operations—not even operations inherited from a parent object—that object serves no purpose.

Conversely, anything users can *do* must be done to *something*—some *object*. There should be no operations that are not assigned to objects. For example, in a conceptual model for a bank ATM, a *view balance* operation would be assigned to the user's **bank account** object.

7.1.4 Assign Operations to the Appropriate Object(S)

Sometimes it is unclear which object an operation applies to. For example, consider the objects and operations for a library application. One operation is searching for a specific **book** (by title, author, etc.). It is tempting to say that this Search operation is an operation on a **book**. But think about it for a minute. What is the Search operation searching? It does *not* search a **book**—that would be a *different* operation, one to look *inside* a specific **book** for a word or phrase. Instead, it searches the library's **catalog** for a specified **book**. That means that the library's **catalog** is an object in the application, with *Search for book* as one of its operations.

However, sometimes it truly is unclear to which object an operation belongs. For example, imagine a conceptual model for an online calendar application. Some calendar applications allow users to create and manage multiple calendars for scheduling different types of events, e.g., personal *versus* work-related. In such an application, what is the object for the operation that creates a new calendar? Is *Create* an operation on the **Calendar** object, or is *Create Calendar* an operation on the **calendar application**? The latter approach is more consistent with the solution to the library catalog problem discussed above. However, many designers would opt for the former approach—making *Create* an operation on Calendar—because software designers and developers often assign operations that create a new object to the object to be created (e.g., `calendar.new()`). Designers can choose to do it however they want, but they should be consistent throughout the conceptual model and the user interface designed from it.

7.1.5 Decide How to Model Similar Objects

If several objects have similar operations and attributes, the conceptual model should reflect that similarity and not treat the objects as unrelated. For example, as we discussed in Chap. 4, a conceptual model for an office calendar application might have several different types of events: meeting, birthday, vacation, personal appointment, etc. These different types of **events** would share many attributes (e.g., name, description, time, date), and operations (e.g., create, view, edit attributes, delete).

Two ways to model similar objects are discussed in Chap. 4. One approach is to include only the generic object-type (e.g., event), and model variants with a *Type* attribute that indicates the type of meeting it is (e.g., Type: birthday) (see Fig. 7.1, top). Alternatively, designers can model variants as specializations (subtypes) of a parent object-type. That conceptual model would have an **Event** type with all the common operations and attributes, and subtypes of Event (e.g., **meeting**, **birthday**), each with whatever extra operations and attributes it needs (see Fig. 7.1, bottom).

How do designers decide between these two ways to model object similarity?

If the similar objects are *so* similar that their operations and attributes are identical or nearly so, it makes more sense to model the subtypes using a *Type* attribute. However, if the similar objects have overlapping but different operations and attributes, it makes more sense to model them as objects that are subtypes of the generic object-type.

A second reason for using object subtypes rather than a *Type* attribute is if designers need to represent sub-categories of any of the subtypes. For example, if a calendar application has **meeting** as one subtype of **event**, and the calendar needs also to provide different types of **meetings**—e.g., **all hands**, **manager-employee one-on-one**, **design team**, etc.— then it is necessary to model the types of **event** as object subtypes rather than as attribute values, so **meeting** can also be broken down into *its* subtypes (either by subtypes or by its own **type** attribute).

7.1.6 Decide Whether to Include and Expose a Parent Object as Well as Its Subtypes

Even if designers decide to model similar objects as subtypes of a parent object-type, they are under no obligation to include the generic object-type in the conceptual model.

Objects	Attributes	Operations
Calendar	title, owner, current focus (day, month, year), events, to-do items	browse, print, create, find event, add/delete event, change focus
Event	name, description, date, time, duration, location, repeat (yes/no), type (meeting, birthday, vacation, personal appointment), invitees (list of persons), reminder	examine, print, edit attributes
To-Do item	name, description, deadline, priority, reminder, status	examine, print, edit attributes, mark complete
Person	name, job-description, office, phone, email	contact (email, phone, text), view/edit attributes

Objects	Attributes	Operations
Calendar	title, owner, current focus (day, month, year), events, to-do items	browse, print, create, find event, add/delete event, change focus
Event (subtypes indented below)	name, description, date, time, duration, location, repeat (yes/no), reminder	examine, print, edit attributes
Meeting	host, invitees, call-in #, access code	announce, cancel
Vacation	start date, end date	
Personal appointment	private?	
Birthday	invitees (list of persons), surprise?	announce, cancel
To-Do item	name, description, deadline, priority, reminder, status	examine, print, edit attributes, mark complete
Person	name, job-description, office, phone, email	contact (email, phone, text), view/edit attributes

Fig. 7.1 Two ways to model Events in a conceptual model for a calendar application

For example, a conceptual model for a home banking application could include **savings accounts, checking accounts**, and **time-deposit accounts** without including the concept of a generic **bank account**. In this example, if designers chose not to include the generic bank account in the model, it would mean that the user interface could never expose generic bank accounts to users.

Similarly, a conceptual model for an office calendar could include **meetings, birthdays**, and **vacations** without including the concept of an **event**. Of course, the application's developers will probably include the generic object-type in their *implementation* model to simplify the coding, but that doesn't mean the generic object has to be exposed to users.

7.1.7   Identify the Attributes of Each Object-Type

An application's conceptual objects are its content—the data users create and manipulate when using the application. Objects contain information in the form of attributes. In the calendar conceptual model described above, **event** has attributes like *date*, *time*, and *location*, and **person** has attributes such as *name*, *job-description*, and *email address*. In a server management application (see Chap. 6), **alarm** would have attributes like *time*, *date*, *severity*, and *status*.

Chapter 3 explained that conceptual models need not and should not describe every detail of an application. If they included all details, they would be the application, not a model of it. A conceptual model should only include sufficient detail to convey to the application's users the ability to predict what they can and cannot do with it. For example, a conceptual model for an e-commerce website could have a **customer** object with *payment information* as a. Payment information could in theory be broken down into several pieces of information—credit/debit, bank, number, expiration date, name, address, etc.—but for conceptual modeling purposes that level of detail may not be necessary. Similarly, a conceptual model for an ATM or an online banking application might have a *preferred language* attribute for each **customer**, but there is probably no need to specify all the languages to be offered.

7.1.8   Decide What Type of Values an Attribute Has

An attribute must have a value. The value has a *type*, e.g., text, number, date, day-of-week, month, etc. For example, Name attributes and Description attributes would have text as their value. A Date attribute would have a date of the year as its value. A bank account's Owner attribute would have a Person or User (i.e., an object) as its value. A meeting Invitees attribute would have a *list of person* objects as its value. Thus, the value of an attribute of an object can itself be an object.

Be careful not to confuse attributes with attribute values. For example, in a server management application, *active* and *resolved* are not attributes; they are possible values of a *status* attribute on the **Alarm** object. In an online clothing store, *small*, *medium*, and *large* are not attributes; they are values of a *size* attribute on the **clothing-item** object.

7.1.9   Watch Out for Hidden Objects

If designers add an attribute and notice that its value is an object that is not in the model, they must either add that object to the model or change the attribute not to have that object as its value. For example, in a conceptual model for a restaurant review website, **user** will be one object. **User** objects will probably have a **favorites list** containing favorite **restaurants**. That implies that the conceptual model needs a **Restaurant** object.

Similarly, if the designers of a calendar application want **events** to have an Invited attribute that is a list of **Persons**, then the conceptual model needs a **Person** object. If the designers don't want to include a **Person** object in the model, they must change the Invited attribute to have as its value something that isn't a list of objects, e.g., a list of email addresses for people not necessarily known to the application.

In general, container objects need an attribute listing the contained objects, but the contained objects must also be included in the conceptual model. For example, one object in a conceptual model for a social media application may be a **posts-feed**. The main attribute of a **posts-feed** is a list of **posts**, implying that the conceptual model must have **post** as an object. The attributes of a **post** would include who posted it, the date and time, number of likes, comments, etc. Similarly, unless it is empty, an e-commerce **shopping cart** contains **products**, implying that **product** must be an object in the model.

Some important objects may only exist as *parts* of other objects. In social media apps, there is no such thing as a **comment** that is not attached to a **post**. In e-commerce websites, **products** have **reviews**.

If an application supports interaction between users, the conceptual model must distinguish between *this* user and *other* users. For example, in a social media app or website, *this* user can edit their own profile but not the profiles of *other* users, and *other* users cannot edit *this* user's profile. A good way to model this distinction is to have two object-types: **this-user** and **general-user**.

7.1.10 Decide How Detailed to Be in Modeling Common Operations

Some operations are covers for collections of related sub-operations. Designers have a choice when creating a conceptual model whether to include only the cover operation, both the cover and the detailed sub-operations, or only the detailed sub-operations.

For example, in a conceptual model for a document editor, documents (an object) might have an Edit operation. But users can edit many things in a document: paragraphs, words, fonts, paragraph styles, margins, etc.

Similarly, in a calendar conceptual model, events (an object) presumably have attributes (e.g., Name, Description, Start Time, End Time, Date, Repeat) that could be modeled by simply giving events an Edit Attributes operation, or it could be modeled by providing Edit Name, Edit Description, Edit Start Time, etc.

How designers should handle such a situation depends on how detailed the conceptual model needs to be for those working *from* it—e.g., programmers, technical writers, and marketing copywriters—to be able to understand and use it effectively.

7.1.11 Include All Task-Relevant Operations

Any task-domain operation that the application will support should be included in the objects/operations analysis. This includes operations that may not require any user action other than looking at the user interface.

For example, getting the current time of day from a mobile phone usually requires nothing more than glancing at the time display. Similarly, mobile phones, office phones, and answering machines often have an indicator that shows if new messages have been left. Determining if any messages have been left requires no clicks or commands—only glancing at the phone. Nonetheless, it is an operation that should be included in the device's conceptual model. One could imagine *other* devices that required pressing buttons or calling a message-center to check for messages.

Similarly, most personal computer operating systems provide a way for users to find out what time of day it is. In Unix and Linux systems operated via command-line shells, users issue an explicit command to do this. In most window-based operating systems, users need only move their eyes to where the time is continuously displayed on the screen. The UI for checking the time happens to be a "no-click" UI (Isaacs and Walendowski 2001). Conceptual models are independent of any specific user interface, so even though it may have a zero-click UI, checking the time of day is a user operation, so a conceptual model for an OS should include it as an operation.

7.1.12 Remove Obsolete Concepts

Sometimes new technology automates a function that formerly required explicit user action, thereby eliminating the need for people to invoke or control it. In such cases, software designers can simplify the application by eliminating the corresponding concepts from the application's conceptual model and hence from its user interface.

For example, early word-processing systems performed spell-checking on demand rather than constantly. When manually invoked, it checked spelling throughout the document and marked misspelled words. It was designed that way because computers and word-processing software were not fast enough to check spelling in real-time while keeping up with users' editing actions. In a conceptual model for those old word processors, **spellcheck** would be an operation on a **document** object. In modern document editors running on today's computers, spell-checking occurs automatically and constantly in the background unless disabled; it is an ON/OFF setting—i.e., an *attribute*—rather than an explicit *operation*.

A concept that is almost obsolete is the **check**. Fewer and fewer people pay with checks nowadays, and many stores and services no longer accept them. It won't be long before **check** objects disappear entirely from the conceptual model of shopping and banking applications.

When a new application's conceptual model omits concepts that were familiar to people, designers should ensure—through user research—that users are not disoriented by the concept's disappearance. Usually, they won't be. People like things simple; the simpler, the better. For example, other than professional photographers, few people who remember photographic **film** and photo **negatives** miss them.

However, when familiar concepts are removed from a conceptual model, the documentation describing the model—for both developers and users—should explain the absence of the concept, to make clear that the concept was no longer necessary.

7.1.13 Part-of and Containment Relationships Need Not Always Be Distinguished

As mentioned in Chap. 4, even though part-of and containment are usually conceptually distinct relationships, sometimes it is not worth the trouble to distinguish them. If users need not understand the difference, designers can safely treat part-of and containment as the same relationship.

Furthermore, in some situations, they cannot be distinguished. Are grains of sand in an hourglass contained by the hourglass or are they part of it? They are both. Are figures in a book chapter *contained* by the chapter or are they part of it? Again, both.

7.1.14 Include Software Developers Who Have Experience with Object-Oriented Programming

Objects/Operations analysis is similar to the Object-Oriented analysis that programmers do when preparing to write object-oriented software code. Therefore, a way to raise a team's skill at performing Objects/Operations analysis for conceptual modeling is to include people on the team who have experience doing Object-Oriented analysis for implementing software. But such team members, if present, must be careful not to add objects to the conceptual model that are purely implementational and not meant to be exposed to users.

7.2   Supporting Learning

Regardless of the subject matter of an application, its users will need to learn its conceptual model, at least partially. The more completely they learn it, and the more accurately their mental model matches the designers' intended conceptual model, the better. How can designers construct a conceptual model to facilitate learning the application?

7.2.1 Metaphors

As discussed in Chap. 4, a conceptual model can be designed to reflect or reference another (presumably well-known) conceptual model. The referenced conceptual model is often called a "metaphor". For example, the personal computer *desktop* with its *files* and *folders* is a conceptual model built using as a metaphor the conceptual model of a physical paper-based office with its *desktops* and *files* and *folders*. (The electronic objects are said to be represented by *icons*, which is *itself* a metaphor based on religious imagery.) E-mail is based on physical memos, with their subjects, dates, to- and cc-addresses. Spreadsheets have arrays of cells, analogous to the cells of paper spreadsheet. Letting one conceptual model stand on the shoulders of another is a powerful way to accelerate learning and provide support for use.

However, analogies and metaphors are limited and inexact (Halasz and Moran 1982; Johnson 1985, 1987). Users may be misled into expectations that the application does not fulfill. For example, the amount of material in a real-world physical folder is usefully indicated by the folder's size and weight. In contrast, the amount of material in an electronic folder is *not* indicated by the size of the folder's icon on the electronic desktop, and the icon has nothing analogous to weight, e.g., dragging the folder doesn't become more sluggish the more it contains.

Designers must take care in conceptual design to ensure that the objects, attributes, and operations of the reference conceptual model (the metaphor) are carried over into the application's conceptual model. Where they aren't, users must be made aware of the discrepancies.

7.2.2 Consistent Terminology

Another way to support learning is to have a consistent terminology. As we mentioned in Chap. 4, objects, attributes, and operations in a conceptual model need not be named using the exact terms that will be used in the final product, especially at first. Over the course of design and development, the concepts in the model and how they are named will probably evolve. However, at any point during a project, concepts should be named consistently so everyone understands each other. Especially in the final product, concept naming in the application and its documentation should follow the following rule to avoid confusing the application's users:

Same name, same thing; different name, different thing (Jarrett 2008).

7.3   Conceptual Model Versus User Interface

Sometimes it is difficult to determine whether an issue is the concern of the conceptual model or of the user interface. Two common but difficult cases are described below.

7.3.1   UI Terms Versus Conceptual Terms

The conceptual model is different from the user interface; the user interface exposes the conceptual model to users. Anything concerning UI details should be left to the user interface and should not be in the conceptual model.

However, the line between conceptual model and user interface can be fuzzy. For example, most calendar applications allow users to choose whether the calendar presents events for one day, one week, or one month; this is often called selecting a calendar "view". It may seem that the view is about presentation and so should be part of the user interface, not part of the conceptual model.

In fact, the underlying *concept* is fine for inclusion in a conceptual model, but the word "view" is just a poor name for the concept. Digital calendars contain too many **events** for people to see at once, so they must provide a way for people to adjust the *scope* of the dates the calendar presents. The **focus** could be based on a variety of criteria: the desired time period (day, week, month, year), which time period (this week, next week, October, November 2013), the type of event (meetings, birthdays), the people involved (me, Aunt Agnes), etc. The **focus** can affect not only which **events** are presented, but also which are acted upon (e.g., deleted). Thus, instead of having a *view*, a calendar can be (re)conceptualized as having a *focus* attribute that controls what dates or events are available for users to examine or edit.

In cases like this, designers should look for concepts that are hidden or mis-cast by sloppy labeling. A way to test this is to determine whether the same conceptual model can be applied for a variety of user interfaces differing radically in their presentation. For the calendar, this would require checking to see if the conceptual model works for user interfaces that do not support visual "viewing", such as Amazon's Alexa, or user interfaces for sight-impaired users. The concept of **view** does not make sense for non-visual user interfaces, but the more general concept of **focus** makes sense for any user interface.

Similarly, applications that can display large amounts of information often have settings to control how much information is displayed. Conceptual models for such applications should use conceptual-level terms such as "filter" rather than presentation-level terms like "view".

7.3.2   Interactive Concepts

Another challenging case of separating the conceptual model from the user interface occurs when an application is being designed to support tasks from a task domain that is fundamentally interactive: visual, auditory, or tactile.

For example, the task domains of graphical information system (GIS) applications include manipulation of maps, and maps include concepts that are typically considered part of the user interface: regions, locations, colors, and symbols. Similarly, Web design tools and document editors manipulate layout concepts (e.g., inline, floating) and distances (absolute: 200 pixels; relative: 25%).

The concepts in a conceptual model should be those that can be manipulated through *any* sort of user interface. For example, maps might have insets, and insets can be conceived of as having a *scale* attribute to provide for different amounts of detail in the same map. Of course, a user interface for maps having such a conceptual model would have to provide ways to *set-the-scale* on **insets**, which could be achieved with a two-finger swipe on a touchpad.

7.4   Object Identity

In applications that share objects between locations or users, the conceptual model should make it clear whether the sharing is achieved via copying or linking the objects. This issue arises in several different situations.

7.4.1   Containment

A common but often blurred distinction in conceptual models occurs when objects are contained in containers: can a single object be in multiple containers at once?

For example, in Apple's e-mail application, Mail, *messages* can only be in one **folder** at a time. Operations on mail **folders** include *move message to folder* and *delete message*. In contrast, Apple's Photos application allows **photos** to be in many **albums** at once, and operations on **albums** include *add photo* and *remove photo*. Importantly, deleting a photo from an album does not remove it from other albums in which it has been placed. Photos also has a **library**, which contains all the **photos** and has its own *deleting photo* operation, which *removes* the **photo** from the **library** *and* all **albums** that contained the photo.

A conceptual model in which objects can be in multiple containers at once is needed only when objects can change. For example, if *photos* cannot change, then sharing can be implemented by copying them. The copies will all be identical and will remain so, and therefore it doesn't matter whether they are identical or copies. However, in Apple's Photos, *photos* can change: they can be annotated, cropped, and adjusted. Conceptually, changes are on the *photo* everywhere it appears, not on the instance of a photo in an *album*.

When change and containment interact, designers should take care to ensure that the conceptual model is clear.

7.4.2   Synchronizing Objects

A recent variant of containment is seen in systems that move and share objects across multiple computing devices. For example, people now expect calendar applications to create events that can be shared between different calendars. Similarly, the Dropbox and SugarSync services support sharing file system hierarchies.

However, the conceptual models for these applications reflect the fact that copies are being made and synchronized with each other, and that it takes time to move events and files between computers. These new models deal with recognizing intended *identity*, *change*, *content*, and *time*. A central idea in them is that human concurrent activity on computers that are intermittently connected may result in conflicts that require human intervention to resolve. For example, applications that manage calendars, contacts, and even files (e.g., DropBox, Google Docs) can merge conflicting changes made simultaneously by people using different computers.

7.4.3   Inheriting Attributes

In complex applications, objects may occur in hierarchies, e.g., type or containment. An object's attributes can be inherited from parent objects. An issue arises when users edit an inherited attribute.

For example, imagine an application for architects to use in designing university campuses. A campus has buildings and buildings have rooms. Buildings have attributes such as *name*, *building number*, *location*, etc. A building may also specify a *color*, which applies to all its rooms. Rooms have attributes such as *room-number*, *capacity*, *phone number*, and *color*. If a room's color is changed, does that affect the color of all rooms?

This is a conceptual design issue, and so must be made clear in the application's conceptual model; otherwise, many users will misunderstand the model and change data that they did not intend to change or will be frustrated by not be able to change data they want to change.

7.4.4   What Work is Saved, When? And Can I Reverse It?

When someone edits a document in Microsoft Word, they are editing a copy. They complete their work by invoking "Save" or (rarely) "Cancel". In contrast, in MacOS or Windows, when a person modifies settings in preference panes or moves files around in folders, they are working on the thing itself, not on a copy. They complete their work by simply stopping (e.g., closing the preference window). Similarly in Intuit's Quicken home-banking application, when users add, edit, or delete transactions, the changes are immediately reflected in Quicken's files. In such cases, letting users back out of actions requires recording user actions, developing the technical capability to reverse them, and giving users a *revert* or *undo* operation.

This distinction between these two ways of working matters most when applications fail (e.g., freeze or terminate unexpectedly): is the work a person has done lost because the copy was not saved, or is it preserved because the data was changed as the user worked?

Although this distinction is relevant in almost every conceptual model, there is no accepted language for discussing it, and no common notation is used across applications and platforms to expose it. It is useful as part of conceptual design to make clear which concepts are handled in which ways, and to evaluate whether the conceptual model is coherent. If not, users will certainly make mistakes in how they think about and treat the modification of objects.

Enhanced Conceptual Modeling

Chapter 7 discussed conceptual modeling issues that developers *always* encounter when creating conceptual models. It discussed how to address these issues in order to make good basic conceptual models.

This chapter discusses three additional issues that can push a conceptual design toward greater richness, sometimes at the expense of complexity:

- Using Companion Models,
- Modeling Interactions, and
- Evolving the Application.

While these issues are *common*, being applicable to most applications, they are also usually *optional*; complete designs can be devised without addressing them. Nonetheless, they are *important*. When they *are* addressed, the resulting designs may be more usable and applicable to more domains, tasks, and plans for task performance.

Because these issues are optional, designers may be tempted to postpone considering them until later in development, with the thought that extra complexity can be added to the conceptual model later if desired. However, postponing can make it harder to evolve to a richer conceptual model later. An attractive alternative is to consider these topics early in development to increase the chances that coherent designs will be possible in the future, and that users will be able to upgrade to them. A redeeming grace is that concepts needed to support these topics need not be exposed to users in earlier conceptual models. All that is required is a migration path to these anticipated futures if the application succeeds.

© The Author(s), under exclusive license to Springer Nature Switzerland AG 2024   83
J. Johnson and A. Henderson, *Conceptual Models*, Synthesis Lectures
on Human-Centered Informatics, https://doi.org/10.1007/978-3-031-50852-3_8

8.1   Using Companion Models

Large multi-function applications often have complex conceptual models. The sheer size of the conceptual model can be overwhelming, and not all parts of it may require agreement from the same collection of developers.

8.1.1   Progressive Disclosure

When many users will need to use only a small subset of an application that supports a rich set of tasks, designers often provide more than one user interface for the application. This is called "progressive disclosure", reflecting the idea that as the user's tasks progressively become more complex, the application discloses more of its conceptual model (Johnson et al. 1989; Johnson 2007).

For example, on photocopiers, simple tasks are supported by a small subset of the conceptual model: make one copy, make a small number of copies, or make the copies 2-sided. However, with a photocopier that can shift images on pages, provide covers, collate, staple, and bind, these extra capabilities are often hidden until revealed by opening a covering panel. When the copier controls are on a display screen, "Advanced features" buttons display the more detailed controls.

Progressive disclosure may seem to be a user interface issue rather than a conceptual modeling issue. However, a better way to think of progressive disclosure is that there are *two* conceptual models, one simpler and one richer, with mechanisms for moving between them. In such cases, it is important that the simpler model be presented as a clear and exact subset of the richer model. Therefore, objects and operations from the richer conceptual model should be sharply delimited so that they do not leak into the simpler one. It is particularly hard on users when the richer conceptual model requires that simple tasks be thought of differently than they are in the simpler conceptual model. When this happens, users can easily become confused about *both* conceptual models.

8.1.2   Component Models

One way of addressing the complexity of a large conceptual model is to separate it into components, each of which is modeled more or less separately.

For example, the conceptual model of a car might include component models for operating the car (steering wheel, pedals, etc.), for adjusting the car's amenities (windows, seats, heater, radio, etc.), and for the engine. In the user interface, some or all of the components—and therefore some or all of the component conceptual models—can be hidden until needed behind progressive disclosure mechanisms (e.g., the car's hood, which conceals the engine), or they can all be available constantly.

Another example: A social media app could have one conceptual model for managing profiles and friends lists, and another for reading and responding to posts, since those two types of activities are fairly distinct.

8.1.3 Surrounding Models

All applications function and are used in some computational context or environment. The environment provides a surrounding conceptual model that users may be aware of, which may impact the conceptual models of applications that run *in* that environment. Here are some examples.

Example 1: Time
Most computer systems represent time in a very technical way (e.g., the number of seconds since midnight, Jan 1, 1900). However, an application would represent times-of-day in a more user-friendly and task-dependent way. For instance, a calendar might indicate that a meeting starts at 10:00 am Pacific Daylight Time on a particular day. The conceptual model would include something like:

- a **moment-in-time** is independent of location;
- a **time-zone** connects a **moment-in-time** to a place;
- **time-intervals** (**year, month, week, day, day-of-the-week, hour, minute, second**) depend on **time-zones**.

The application would have to integrate these time-related concepts with its other concepts. For example, in a calendar, although a meeting starts at a moment in time, the time-labels for that start time will differ for participants in different **time-zones**; it may even be on different days or even years. Are users of the application only interested in what the time is in their time-zone, or do they want to know how other participants see it? What time will that be in New York City? Is that a holiday in Bangalore?

Example 2: Files
Many applications borrow concepts from the operating systems in which they run. One of the most common concepts borrowed from the operating system is *files*.

For example, in one company's application for creating **maps**, a key concept in the conceptual model was **table**, the aggregated data defining a map. Historically, the application had used spreadsheet files to represent **tables**, so users could edit them with spreadsheet applications, and the mapping application did not have to support table editing. The designers felt that because these data **files** were accessible via the operating system, and users were familiar with editing data as spreadsheets, the application did not have to provide its own access to them. Users made use of other applications—the operating system and the spreadsheet application—to edit map data. Thus, the mapping application borrowed concepts from other applications in its environment—file management and spreadsheet editing. However, the mapping application did check imported spreadsheet files to ensure that they contained valid map data-tables.

Example 3: Security

In applications that connect to the Internet (and few do not these days), access to objects must be carefully managed. Sometimes, applications run entirely within protected environments (e.g., a secured machine), so security is not an issue for their conceptual model. More often however, Internet applications allow users to access objects remotely. For example, events on a shared calendar may be accessible by some people and not others. As another example, bank accounts are managed in highly secure banking systems, but they usually can be accessed remotely by properly authenticated account owners.

The computation platforms on which applications run provide powerful mechanisms for limiting access, e.g., access control lists on files, encryption on passwords and data transmission. However, the concepts comprising these access-mechanisms are rarely the right ones for describing how security works in the application's domain. For example, access in a calendaring domain should reflect something like:

- access is not given to **login names** or **machines**, but rather to **projects** and **members**.
- access is not on **files**, but on *events* that are contained in *calendars*.
- the possible access attribute values are not *read* and *write* (as for **files**), but rather *show-existence-only, show-title-only, show-details*, controlling how much others can see.

The point of these first three examples of surrounding conceptual models (time, files, security) is that the conceptual model of an application can be augmented by borrowing concepts from the conceptual model(s) of the platform(s) on which it runs. Whether a conceptual model augmented in this way will be easier or harder to learn depends on whether the application's users understand the platform's model.

Example 4: External physical objects

Another important type of concept that is sometimes overlooked in building conceptual models is external objects—real-world objects outside the computer. If an application supports users in manipulating external objects, those objects should be included in the application's conceptual model.

For example, a photocopier carries out jobs in which it scans "original" documents, feeds blank "supply" paper, and produces documents that are visual copies of the original. A conceptual model for a photocopier must include documents and paper. It must also include "areas" of the paper path for clearing paper jams. The copier may also require that toner (ink) cartridges be added periodically, so the conceptual model must include those. Finally, photocopiers also require oiling and other maintenance functions, but only maintenance personnel would need conceptual models containing maintenance tasks and concepts.

8.2   Modeling Interactions

The use of an application can be considered as a series of *interactions*[1]—conceptual operations taking place over time. So-called "history" mechanisms treat interactions as concepts and allow users to navigate through past interactions. History mechanisms are important resources for users, in handling errors, changing direction, and preserving and repeating activity. This raises several issues when designing conceptual models.

8.2.1   Managing Errors

Errors are common when people use applications. Some errors arise from incorrect user mental models: misunderstanding the application's conceptual model. Some errors result from faulty operation of the user interface; such errors are either slips (mis-typing, mis-pointing, mis-speaking) or mistakes (mis-reading, misunderstanding the menus or the displayed information) (Norman 1983; Johnson 2020). When a person makes an error while using an application, they often want to back up, repair the damage, and try again.

Similarly, when people change their plans as their work progresses, they may want to go back and make adjustments before proceeding. For example, a graphics artist might decide to change the colors in a document because their initial color choices are hard to read or do not fit with a new color strategy.

When such changes in direction occur, recovering from them requires doing four things, which should be supported by the application's conceptual model (and delivered through its user interface, of course):

1. *Detect*: Notice that something unexpected has happened, that there is a discrepancy between what one expected and what one detects. Expectation is expressed in terms of a conceptual model; what one encounters is also registered and understood against that conceptual model. A compounding difficulty is that a person may not notice discrepancies until several following interactions have been performed.
2. *Diagnose*: Figure out what happened. The clearer the conceptual model, and the better the access to the state of the application expressed in terms of the conceptual model (of course), the easier is this conceptual work.

---

[1] We use the term "interaction" here to avoid conflict in terminology with our running example of calendar applications whose CMs in this book include the concept of "event".

3. *Repair*: Determine how to repair the problem. Can a user just change things to the intended state, or must they "undo" and then proceed, or do they have to abandon all recent work and start over from the beginning? The provision (or lack thereof) of "undo" capability is important to note for operations in conceptual models (Aboud and Dix 1992).

4. *Resume*: Proceed. However, not as though nothing had happened, but rather with an awareness that non-standard flow has occurred, forcing intervening activity. For example, *undo* is itself an interaction. After undoing an interaction, the history list may well show the last interaction was *undo*. If the user then invokes *undo* again, what will be undone: the *undo* or the previous interaction?

For example, when a person is editing a figure in an electronic document, moving a shape around in the figure may produce unexpected results (e.g., the shape disappears). With effort, the person might determine that the object is underneath another object. They undo the move, bring the object forward, and move it again. They proceed, knowing that the application has *undo*, which can undo an *undo*; *step backward*, which always goes back in time; and *step forward*. And what has the impact of this activity been on resource usage? If operations use resources, did the user expend twice as much? Must they "clean up" (e.g., empty the trash)? The application's conceptual model should make such things clear (Jackson 2021).

Another example comes from the operation of *friending* someone on social media. The act of *unfriending* someone is usually not the same as *undoing* the *friending* operation for that person; the history does record that the pair of operations happened, and the social situation it represents does too.

Sometimes, adding concepts to handle interactions requires fundamental changes in the conceptual model, and thus in how users will have to think about their work. For example:

- *Interactions as objects*: To manage interactions, people must change their focus of attention to the interactions that have occurred in the application. This requires that the application has maintained a record of the interactions carried out, the objects involved, and the sequence in which they occurred. The conceptual model must therefore be enlarged to reflect this, including concepts like invocation (attributes: *operation, arguments, results*) and interaction (attributes: *invocation, prior-interaction, next-interaction, time*).
- *New operations extend the conceptual model*: Undoing needs to be understood in terms of the concepts of the conceptual model. For example, the conceptual model for a bank account, which has a *transfer* operation, could have an *undo-transfer* operation. However, note that the *undo-transfer* operation must be different from *withdraw*: *undo-transfer* must follow a *transfer* (not so *withdraw*), and a *transfer* followed by an *undo-transfer* might leave no record on the bank statement (not so *deposit*, then *withdraw*).

- *Appropriate grain-size:* A well thought-out, multi-interaction *undo* often requires answering difficult questions about the grain-size of the interactions that are recorded and undone. We expect a text editor to respond to each key typed; each keystroke is its own *insert-character* operation. Yet a corresponding *undo* operation that undoes only one *insert-character* has been shown to be much worse than developing some concept of a *typing operation* that reflects chunks of typing as people think of (actually, intuit) it, and providing an *undo-typing* operation (Washizaki and Fukazawa 2002).

One important design consideration is that the conceptual model for history should be broken into two parts, one reflecting the conceptual model of the application, and one reflecting the user interface through which the user is accessing it. Good design will provide all the functionality of the application—including what is needed for managing interactions—through not just one but all user interfaces.

8.2.2 Anticipating Trouble

Users also knowingly choose to save the state of an application's data as insurance against trouble and as support for exploring alternatives and changes in direction. For example, databases provide ways to create checkpoints (for backing up) and group transactions (for making a collection of changes all together or not at all). However, these capabilities are often not reflected in the conceptual model of applications that use those databases.

8.2.3 Macros: Capturing Activity

Work is often repetitive. For example,

- graphic designers test filtering operations by first applying them to one image, then selecting a subset of the filters to apply to many images,
- statistical analysis often requires multiple tests, winnowing of datasets, and shaping of presentations, repeated over multiple datasets,
- text editors are used by writers for managing hierarchies of chapters and sections, each with many versions.

A common mechanism for supporting repeated work is "macros". Macros are user-accessible languages that enable the description of sequences of interactions. For example, Unix provides shell scripts, MacOS provides AppleScript and Automator, and Excel provides Visual Basic.

A common mistake is to have the macro-language access applications using the descriptions of what happens at the application's user interface. This limits the macro-language to work only with that user interface, requiring different macros when a person accesses the application through a different user interface, or the user interface is changed.

A better approach is for the objects and operations of the macro-language to be exactly the objects and operations of the application's conceptual model. For example, **cells** are a key object in Excel's conceptual model; and **cells** are accessible in Visual Basic within Excel. Ideally, the concepts in Apple's Automator are exactly those in the conceptual models of MacOS applications.

8.3   Evolving the Application

Over time, tasks in domains change because, for example, the world changes, better methods are adopted, or better plans are employed. As part of conceptual design, it is therefore worth considering how to support the evolution of an application to support changes in how tasks are done. This includes exploring how tasks in that domain are *likely* to evolve, and, correspondingly, how the application must evolve to support that work. It also includes recognizing that the future may not turn out as anticipated, and therefore exploring how the application can support unanticipated change.

8.3.1   Managed Growth

Most applications are created iteratively, particularly in this age of "agile" development methods. As an application grows in functionality, its conceptual model grows with it. In fact, the best way to grow applications is by evolving the conceptual model and then reflecting it in everything else: user interface, documentation, and implementation.

When developers predict the structure of future conceptual models, they should take care to design migration paths that will help users make the necessary conceptual shifts. For example, if a planned calendar application offers **overlays** in a future release, the current version might be designed as if it had a single (perhaps implicit) **overlay**. The impact of that could be that the concept of overlays would be familiar when it is introduced.

8.3.2   Anticipated Growth

Even for designers, crystal balls are unreliable. It is impossible to know what the future will bring. However, it does not take a crystal ball to anticipate that change *will* come, and that designers probably can't predict exactly what it will be.

An excellent strategy is to turn to those who will know the future when they see it—the users. Provide ways for the people using the application to extend it to support the work they do. There are three ways to do this: construction sets, platforms, or extensibility.

Construction Sets

Some applications provide objects—often called "pieces"—that are intended to be combined within the application to create composites pieces. Such applications are often called "construction sets", reflecting construction provided in some mechanical toys (e.g., Lego, Erector Sets, Tinker Toys, and Lincoln Logs). For example, applications for creating illustrations provide a set of **primitive shapes** and a *group* operation for aggregating shapes to create **composite shapes**. Often composite pieces may themselves be further combined as members of even larger composite pieces; this is often conceptualized as **pieces** with the attribute *type* (value: base/composite), with the operations for making combinations working on **pieces** of all *types*. This is called "recursive" or "injective" composition. For example, most applications supporting illustration (e.g., PowerPoint, Illustrator, Omnigraffle, and Pages) have injective composition.

Platforms

People commonly use concepts in one application to *represent* concepts in another. For example, in the domain of financial management, the conceptual model of a spreadsheet application has *cells* and *formulas* as objects. However, a person might use a spreadsheet to represent their monthly budgeting task. The concepts of budgeting would be represented in the spreadsheet, say with columns for *budgeting categories* and *amounts*. Further, they might copy last month's budget (spreadsheet) to start this month's budget. Over time, they might come to know what parts of the spreadsheet change, and what parts remain constant, month to month.

For them, the resulting spreadsheet is both a spreadsheet and a *budget*. When one task-domain is used to represent another, we say that the application used to do the representing is being used as a "platform". Spreadsheets are intended to be used as platforms. Other programs that are not so intended are *also* used as platforms: a text-editing application is used as a platform for invoicing; a calendar application is used as a platform for scheduling; an e-mail application is used for fund-raising.

When designing an application that is intended to be used as a platform, the tasks are considered at two levels: the domain of doing representations (using the application to represent other applications), and the domains that can be represented. The concepts of platform are a surrounding model (see *Surrounding models*, above) for the concepts created using the platform. The conceptual model of the application should include concepts that make the work of representing easier. For example, spreadsheet programs have rows and columns; they also have a *make-chart* operation that takes a data range, treats it as representational (data must be in a particular form), and creates a chart.

Note that the conceptual model of the represented application is usually created by the user of the platform application. That is, the user of the platform is acting in the role of the designer of the conceptual model. For example, as mentioned, a finance application might have a conceptual model with objects of type *budgeting categories* and *amounts*. To reflect that fact, the user of the representing spreadsheet application, acting as the designer of the represented budgeting application, chooses to label some columns *budgeting categories* and *amounts*. However, notice that there is no requirement that the conceptual model of the represented application be made explicit. Also note that the user is not expected to be a good conceptual designer of the represented spreadsheet application, a fact that many large institutions that use institutionalized spreadsheets for their accounting learn at their cost.

In inputting data that came from a representing application, but was captured in some other format, an application can appeal to users for help. For example, in a spreadsheet application, when textual data that represents a sourcing spreadsheet is being read, users had to be asked for help in indicating where the rows and columns appear in the representing text.

Extensibility

Applications can provide for concepts to be created, not at design time by designers, but *during use* by users. For example, many text-processing applications allow users to put headings of various types into documents. These built-in *paragraph styles* reflect the standard practices of typographers. Because these practices are not universal, many text-processing applications also let users create their *own* styles to reflect their typographical practices. Thus, the application is designed to be extended: the concept of *style* is extensible.

The conceptual mechanisms for describing extensibility include making choices (e.g., selecting a value for an attribute), composition (e.g., aggregating choices in paragraph styles), and specialization and customization (restricting an attribute's range of values). The technical mechanisms for implementing extensibility, however, can include things that are not within the conceptual model of the application at all. In particular, many applications permit extension through programming (e.g., scripting) in either application-specific formalisms or general-purpose programming languages. For example, in describing webpages, JavaScript can be used to extend the behavior of the foundational HTML forms.

Managing Versions

As applications change, new versions are made and released. These new versions may have new conceptual models. Both designers and users expect the new versions to be "backward compatible", so data produced from the older version is acceptable to the newer version. Therefore, an application must somehow mark the objects that it produces to indicate which version of the application produced them. The concepts associated with later versions of the application and their conceptual models are often missing from earlier versions, causing "forward incompatibility".

As new versions become available, people may decide to start using them. People rarely upgrade all at once, so various versions of applications are often in use concurrently. Also, users share their work, and expect their results to work together. This requires that applications and their conceptual models address the fact that objects that are imported may come from older versions of an application and reflect an older conceptual model. When this happens, not only must the implementations be converted, but so must the concepts.

For example, when Microsoft Office moved to XML-based documents, the whole model of embedded graphics in documents changed. Old documents had "graphics frames" (aka "canvasses") in which graphical objects could be drawn, and graphical objects could not be drawn outside of graphics frames. New documents have graphical objects that can be overlaid on anything, and the graphics frames are gone. A user who understood the old model may not understand the new one.

In addition, a new version of an application can appeal to users for help when it detects data from older versions. For example, in a modern calendaring application, meetings account for the possibility that participants are in different time zones. If events are imported into the new version from older versions in which meetings were assumed to be in one time zone, the new version can ask users to manually supply the time-zone information.

What happens when an object being imported was created by someone using a *newer* version of the application with a newer conceptual model? That is, the application is trying to handle something that comes from its future. In this situation, some applications crash (bad); some refuse to proceed (not great); some strip out what they do not understand (better); some preserve and protect what they do not understand and pass it back out undamaged to others to use (best).[2]

Supporting data-interchange between different versions of applications with different conceptual models requires careful thought and planning. It must be addressed in the conceptual model. Specifically, operations are needed to make the required conversions. The newer conceptual models may need to include descriptions from the older ones, and versions of the concepts need to be named to remove ambiguity. One solution is to tag concept labels with version numbers. This is not a solved problem.

8.3.3   Unanticipated Growth

Despite all efforts to analyze the tasks in a domain, it may not be possible to guarantee that the concepts in an application will align well with the actual situations users encounter in the future. Sometimes, this results from a lack of development resources to see or consider

---

[2] This is called "round-tripping".

every task in the domain. More often, the world changes, leaving some misalignment between tasks and the application.

For example, many years ago, a Xerox photocopier was on an ocean-going barge. When the customer tried to order supplies, the order-entry clerk asked for a shipping address, but the customer could not supply one, as the barge was moving between ports-of-call. However, the form that the order-entry clerk was completing required a conventional street address, so the order could not be placed.

When the tasks and the application cannot be made to align, what is the user to do? Indeed, what can designers do to address novel task requirements? How should the conceptual model reflect this inevitable result of working in a dynamic world? The following are possibilities.

Backchannel

Make it possible for users to tell the designer about the difficulty. While this will not solve the immediate problem, it may help in getting a fixed version released. For example, when applications fail (crash!), the implementations often offer to send a message to the implementers providing the technical details of the failure. Similarly, when the design fails, some applications support the user in sending a message to the designer giving the task circumstances. This "backchannel" is the means of getting a message "back" to the designer. To do this, the conceptual model must include concepts for communicating, addresses for designers, and descriptions of task circumstances (e.g., open-ended text, photos).

Tailoring

Make it possible for users to fix the application: to modify it to be as they would like it to be. This throws users into the design process, continuing the design in use. It is fraught with difficulty, because it requires that users be able to not only understand and change the design, but also be able to implement it. The impact on the conceptual model is to include all the concepts of designing and implementing. (Extensibility, discussed above, is a special case.)

Even if all that is possible, the user may not want to switch to doing the application developer's work while in the middle of working on a task. Also, the timescale is probably wrong, since development work usually takes much longer than application usage. For example, when in the midst of writing something using a text-editing application, if a paragraph style has the wrong "next paragraph" setting, a user is not likely to edit the style at the expense of losing their writing flow (Henderson and Kyng 1991).

Appropriation

Find someone who has seen this problem before and has solved it and appropriate their solution. This is a great approach, as it leads toward workable, even working, solutions. However, it requires that people be able to characterize their problem, know enough about other people's activity to know about earlier difficulties, find them, and get their help. Again, it usually cannot be done quickly enough to suit the circumstances. To support this approach, add **users** and **completed tasks** to the conceptual model.

Workarounds

Augment the conceptual model of the application with some objects and operations allowing a user to create a solution that is partly *inside* the application and partly *outside*. This is common practice. For example, the Xerox order-entry clerk (see above) handled the "no available mailing address" problem by entering into the form's address field the text, "Call Bob" with a telephone number. In the paper world, business forms (the UI of the application) have *margins* on them; these are the normal places for everything in the task that doesn't fit the structure of the form (the conceptual model of the application) [Yates, *Control through Communication*]. To enable this solution, the conceptual model is augmented with open-ended ways for recording **exceptions** of one type or another (e.g., illegal values in fields, additional information, and challenges to the assumptions of the application), together with operations for setting them and searching them.

Special Handling

Provide a special way to deal with exceptional cases. This is a mindset that is most relevant in situations where the only way of getting work done is through the application (e.g., all hiring must be done through the HR++ application). For example, InterMountain Healthcare requires that medical staff at their clinics treat patients either by following the established medical protocol (the usual case) or by doing what they deem to be better and documenting it (the special case). The two "lanes" are designed together for coherence: the medical treatment protocol can be rigid (to achieve compliance), and the exceptions drive the constant work of a protocol development committee (to respond to a changing world). Adding special handling to the conceptual model for an application requires concepts like *is special* (on cases), *special handling report*, and *status* (for tracking).

User-Created Conceptual Model

Ideally, of course, it would be great if a user could engage and modify the application at the level of the conceptual model. (This can be seen as the most extreme example of "Tailoring"; see above.) The user would access the conceptual model with an "editing" UI, modify the conceptual model to fit their new needs, and let the application adjust everything to make the application reflect the modified conceptual model. This automatic adjustment might let the user provide some help on how they wanted the changes to be presented by their normal user interface, on aspects of the implementation (e.g., how much file space is the user willing to devote to storing the new objects added), and control on sharing the modified application and UI with other users.

As of this writing, concept-based application evolution by users remains a research topic. User Interface Management Systems (UIMS) focused on generating user interfaces.

Content Management Systems and XForms cover forms-based applications. Model-based architectures, including UML, address object-oriented architectures. Missing are modeling languages that can adequately express the semantics of the conceptual models, and user interface generation formalisms and mechanisms that take into account the technologies that deliver both the application (e.g., running on multiple servers) and the user interface (e.g., often running on a phone, tablet, or other mobile device).

We therefore encourage the research community to revisit the (semi-)automatic generation of applications from conceptual models.

8.3.4   Embedding in Social Domains

The evolution of software applications is rarely an individual matter. Most task domains are collaborative; people's changes can affect others. Each person responds to their own view of the situation. Soon the work can be riding off at high speed in many directions.

Therefore, changes must be considered in the context of the collaborative (cooperative and competing) activity of the task domain (or domains, since activity often crosscuts and interacts). People negotiate, contest, support, and struggle. Most domains are social domains; the social activity encompasses the activity of the tasks and the applications.

As with the methods of anticipating the unanticipated (see *workarounds* and *special handling*, above), applications can be designed to also support the social aspects of the activity that encompasses the task activity they are built to support (see Henderson 2008). That has implications for their conceptual models.

For example, the intense activity that accompanies closing an organization's books at the end of the quarter or the fiscal year is a highly negotiated set of conversations. Financial applications address the anticipated final financial decisions; they rarely address—indeed, they often deny the existence of—the conversations that lead to those decisions. Therefore, every manager in every business invents their own way of supporting these conversations. It might help to extend the conceptual model of financial programs to support all those sticky notes and e-mails, and to support the trading that goes on between managers in order to really do the work well.

Perhaps support for conversations, and the conceptual models and concepts that support them, should be a critical part of all applications. Evolution may not be so optional after all.

Process of Designing with Conceptual Models

9

For any real-world application that embodies more than a handful of concepts, conceptual design can be expected to take a development team at least two weeks of meetings, drafts, discussions, and revisions. This chapter describes how to do this most efficiently and effectively.

9.1 Overview of the Process

Software design and development, when properly centered on the intended users and tasks, includes many different types of work:

- gather and prioritize requirements, including user analysis and task analysis
- conceptual design
- user interface design, prototyping, and implementation architecture
- pre-implementation evaluation: test prototype designs
- implementation
- post-implementation evaluation: usability testing
- documentation (internal and external, online and printed)
- release
- support.

Although the activities are listed here in a logical order, the listed order is *not* necessarily the order in which these activities occur. Later in this chapter, we explain that design is almost always iterative. Elements of *all* of the above activities occur and recur through

© The Author(s), under exclusive license to Springer Nature Switzerland AG 2024   97
J. Johnson and A. Henderson, *Conceptual Models*, Synthesis Lectures
on Human-Centered Informatics, https://doi.org/10.1007/978-3-031-50852-3_9

development, regardless of whether that is planned. For example, testing a prototype user interface may expose the need to return to previous steps.

9.2   Start with User Research to Understand Users and Tasks

A conceptual model for any substantial application cannot be designed out of thin air; it must be based on a good understanding of the users and tasks that the application is intended to support. Therefore, before starting to design an application, design team members should conduct user research: *observe and interview people doing the tasks* (Beyer and Holtzblatt 1997). The resulting data are used to construct user profiles and a task analysis.

9.2.1   User Profiles

User profiles are brief descriptions of relevant characteristics of the intended users of the application. An application will almost always have more than one type of user. For example, users of a server management console might include junior system administrators, senior ones, system architects, and IT managers. Even relatively simple consumer applications or websites will have both new and experienced users.

A profile is created for each type of user, indicating duties, level of education, knowledge of the task-domain (high, medium, low), experience with previous versions or competitors of the application (high, medium, low), knowledge of computers in general (high, medium, low), and any other relevant attributes. Sidebar 3 gives an example of user profiles.

Sidebar 3: User Profiles for a Protein Analysis Instrument

Chapter 3 described a case in which one of us was helping a company design software to control an instrument for analyzing protein interactions. The target market was university and industrial biology labs. In early meetings, the company's design team assumed that all potential users were the same: biologists. However, a bit of user research revealed that biology labs typically have *four different* types of staff who might use the instrument: lab technicians, graduate students, scientists, and power-user scientists. After user interviews, observation, and discussion, the following four user profiles emerged:

User Profile 1: Lab Technician

- Responsibilities: Calibrate, operate, and maintain the instrument. Run (complex) experiments for scientists.

- Knowledge of instrument: high
- Knowledge of biological analysis: low
- Knowledge of competitor instruments: no
- Computer skills: high.

User Profile 2: Grad Student

- Responsibilities: Calibrate, operate, and maintain the instrument. Run simple experiments for scientists.
- Knowledge of instrument: low-medium
- Knowledge of biological analysis: low-medium
- Knowledge of competitor instruments: low
- Computer skills: medium–high.

User Profile 3: Scientist

- Responsibilities: May be the principal investigator. Operate and maintain the instrument. Run experiments.
- Knowledge of instrument: low-medium
- Knowledge of biological analysis: high
- Knowledge of competitor instruments: varies
- Computer skills: medium.

User Profile 4: Power-User Scientist

- Responsibilities: Calibrate, operate, and maintain the instrument. Run (complex) experiments for other scientists.
- Knowledge of instrument: high
- Knowledge of biological analysis: high
- Knowledge of competitor instruments: high
- Computer skills: medium.

Some design teams create comprehensive user profiles that include home-life details, back-stories, and even fictional pictures and names; such profiles are called *personas* (Cooper 2004). Creating personas rather than simple user-type profiles can be useful to design teams that no longer have access to real users after the initial user research. Figure 9.1 shows a persona developed for designing applications aimed at older adults.

Wong, 70 and his wife live in a small apartment near their son, daughter-in-law, and grandson in a small city in China. He enjoys playing Xiang qi (Chinese chess), practicing Qigong, and spending time with his family. Wong uses a desktop computer to keep track

Fig. 9.1 Persona used to design software for older adults (from Johnson and Finn 2017, used with permission)

of his appointments, reports, news, and financial transactions. He uses his smartphone mostly for phone calls but also for weather, transportation, and social networking. He tried a Xiang qi app on his smartphone but had trouble seeing and manipulating the very small objects on the screen.

9.2.2 Task Analysis

In addition to understanding the users, application designers need to understand the tasks that the application is intended to support before constructing a conceptual model. Task analysis starts with observing people working in a domain. Project members engage user-types: observe them, listen to them, interview them, invite them to participate in focus groups, give them surveys, read their user documentation, and talk to their management. These engagements discover not only what tasks people need to do, but what applications could be created to support those tasks.

It is common to listen to the terms that people use while performing or describing their work. These terms suggest concepts for conceptual models for applications. For example, in the banking domain, hearing that people want checking, savings, and investment accounts suggests these as concepts for a banking application (Blandford 2008; Blandford et al. 2013).

There are several ways to represent the results of task analysis. Often, it is useful to represent the task analysis results in more than one way:

- *Major tasks or goals (use-cases).* This is a list of all the high-level goals that users could have in mind when they prepare to do some work. For example, in the task-domain of business accounting, use-cases could include the following:
  - Prepare quarterly statements: profit/loss and balance sheet
  - Prepare end-of-year statements: profit/loss, balance sheet, tax
  - Enter transactions
  - View a transaction
  - Edit a transaction
  - Audit an account.
- Task hierarchy. This is an enumeration—usually in outline form—of all high-level tasks decomposed into their component sub-tasks, including alternative methods of doing the task. It includes breaking sub-tasks down into smaller sub-tasks, etc. For example, editing a transaction requires finding it. A transaction can be found by browsing or by searching. Once found, some of its attributes are edited and (implicitly or explicitly) saved.
- Consolidated task-sequence model. This is a chart illustrating typical sequences of steps to perform a task, consolidated across all observed users (Beyer and Holtzblatt 1997, 2016). Each sequence begins with whatever motivates or triggers it, lists all typical steps including decision points and alternate paths, and ends with the accomplishment (or abandonment) of the initial goal. If different types of users perform the task, the chart would indicate the user-type it corresponds to. As a simple example, a company accountant might notice that one of the accounts doesn't balance, requiring logging in, searching for missing or erroneous transactions, correcting any errors found,

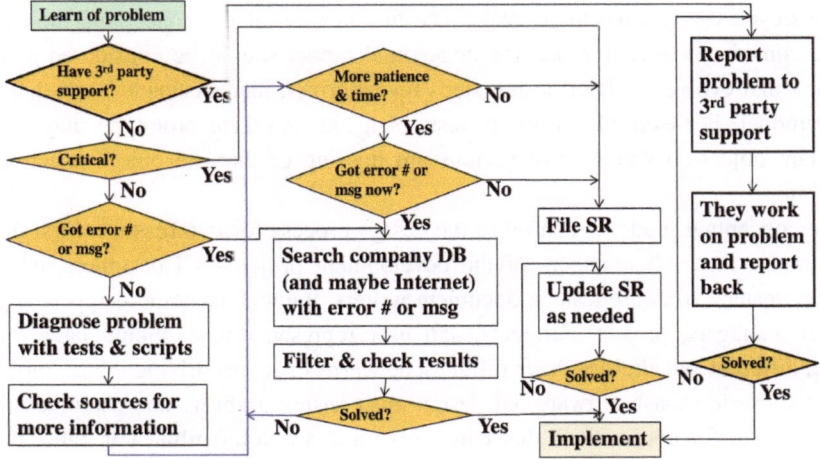

Fig. 9.2 Consolidated task-sequence model for a server system administrator's response to a server problem

rechecking the balance, and logging off. Figure 9.2 shows a consolidated task-sequence model for how a typical server system administrator addresses a server problem.

9.3   Based on User Research, Start Designing a Conceptual Model

With user-research results as input, the design team can begin developing a conceptual model for the application. Chapter 4 describes the important parts of a conceptual model and how to create them, Chap. 5 shows several different ways to represent the Objects/ Operations structure of conceptual models, Chap. 6 presents two realistic examples of conceptual models, and Chap. 7 provides tips on how to design good conceptual models and avoid common mistakes. Before reading on, consider reviewing those chapters to refresh your memory.

9.3.1   The Conceptual Model is Central to the Design Process

Figure 9.1 should be familiar. It appeared in Chap. 2, illustrating the activities involved in software design and development. The process is depicted as a cycle, because release leads to the application in use, with support supporting, which leads to a better understanding of user needs, and to the start of the next cycle. Sometimes, this cycle is ponderous; sometimes it is agile. Sometimes these activities are sequential, but more often they are concurrent.

The arrows from activities to and from the conceptual model indicate that the process contains cycles *within* the main cycle. The first version of the conceptual model should not be considered final. Instead, the conceptual model should be considered a work-in-progress that evolves as later design steps reveal errors, inconsistencies, and deficiencies in the model. For example, usability tests of a user-interface prototype may show that important object attributes or operations are missing or that the object structure needs adjusting.

The conceptual model is central to the design process for two reasons. First, it should be a *team* effort. All members of the development project—UI designers, developers, usability testers, software testers, document writers, trainers, customer support engineers, product managers, project architects, and user representatives—should have input. Of course, different stakeholders have different concerns or forces driving them: implementation must create robust software, UI design must ensure usability and usefulness, project architects aim for applications that can evolve and are compatible with other company products, support must try to keep customers happy, product managers want a timely and successful launch and a successful product, and so on. These divergent interests can cause tensions as the conceptual model is being designed.

A second reason the conceptual model is central is that it communicates the underlying design to all stakeholders. After the team has a conceptual model that everyone can accept, team members should commit to following it in their assigned parts of the project. The project manager or the product architect can check that the current model is adhered to—until it changes.

9.3.2 Coordination is Required

The disparate pressures on development must be managed and coordinated. When conflicts arise, they must be negotiated to produce a shipping application. Often these negotiations can produce novel design solutions where everyone wins. Changing designs may impact several stakeholders.

The methods for negotiating vary widely across projects of different sizes and with different styles of management. They can be formal or informal, may involve documents or talk over lunch, may move between formally adjacent activities, or may shortcut the formal structure entirely.

Whatever development process is used, it is valuable to recognize that this negotiation must take place. A good development process therefore provides a place and occasions for this negotiation to happen. Metaphorically, this can be seen as a table around which all the activities sit and discuss their realities, negotiate, and create solutions.

Our view is that agreement on and changes in the conceptual model are crucial topics for those representing the various roles gathered at this development "table".

9.3.3 One Team Member Should Drive the Conceptual Design

The entire team should have input into the conceptual model, but designing by committee can waste a lot of time. One person—probably the product architect or the lead UI designer—should oversee driving the objects/operations analysis. Based on whatever prior information they have (e.g., marketing research results, user research results, and task analysis), they may want to draft a first-draft objects/operations analysis and present it in a team-meeting for feedback.

Alternatively, the person in charge of the conceptual model may prefer to ensure team members' buy-in by holding a team brainstorming session to produce a first-draft analysis. However the process is started, the best way to proceed is by producing successive drafts and presenting them to the team for feedback.

9.3.4   Include Developers, but Keep the Conceptual Model Focused on User-Facing Objects and Tasks

Because of the similarity between object-oriented software design and the Objects/ Operations analysis that is part of building a conceptual model, it helps to have OO programmers involved. They understand, for example, what it means to attach attributes and operations to objects, and how objects can be related to other objects. They can help inexperienced team members decide what is an object, an operation, or an attribute, and how the concepts are related.

However, software developers should not be the ultimate deciders of what concepts are in the conceptual model, because they often want to expose concepts that should not be exposed: those that reflect the implementation rather than users' conceptual objects, tasks, and goals.

9.4   Use the Conceptual Model to Coordinate Development

The conceptual model should be a central coordination point for development team members and their respective contributions: almost everyone on the team should orient toward the conceptual model as they design and develop the application.

The centrality of the conceptual model and its potential role in orchestrating the design process has one very strong implication for design activities and their relationship with the conceptual model:

> Changes to the conceptual model are *team* decisions. Unilateral addition of concepts to the conceptual model by any team member is not allowed.

For example, if a programmer thinks a new concept needs to be added to the software and exposed to users, she must first persuade the team to add the concept to the conceptual model; only then can it appear in the software. Similarly, if a documenter discovers the need to introduce an additional concept to explain how to use the application, that change must first be reflected in the conceptual model, with the team's buy-in; then it may appear in the documentation.

When a change is made unilaterally, it can seem relatively inexpensive to the person proposing it. However, when all the impacts of a proposed design change are revealed by considering their impact on the conceptual model and the resulting impact on everyone else, the full cost can be considered before committing to the change or deciding against it.

9.5   Iterate, Iterate, Iterate

People new to the idea of conceptual models often see them as an extra step in a linear, "waterfall" development process, e.g., determine requirements, ***develop conceptual model***, design user interface, code user interface and back-end software, test, and release. In such a linear process, once each step is complete, its products—documents, code, etc.—are final and never revisited or revised, and later steps are based strictly on earlier ones.

This view is at the root of many current-day software managers' resistance to developing a conceptual model. Today's software managers often prefer to follow a development process that includes iteration and refinement, and that does not rely on big up-front design efforts and specification documents. This approach is based upon the idea that the initial conceptual model won't be completely right, so a user interface based on it will also be wrong.

It is true that first attempts at conceptual models are often not quite right. It is tough to get it right the first—or even the fifth—time. However, it is *not* true that developing a conceptual model rules out iterative design and development.

In fact, the process is *rarely* linear—even when organizations *try* to follow a "waterfall" development process. As design proceeds from conceptual model to user interface to implementation, it is likely that downstream design and implementation, not to mention user-testing, will expose problems in the conceptual model, indicating the need for changes: additional objects, new attributes for objects, or new operations on objects.

In Fig. 9.3 (above), feedback loops through the conceptual model provide places where new understanding and evaluation findings can cause a return to a prior stage to revisit and revise that stage's output. Early usability testing can, and should, be designed to accelerate this process: low fidelity, quick prototypes can be focused on the important parts of, and questions in, the conceptual model. Lightweight usability testing can thus evaluate the conceptual model as well as the UI design.

In particular, a conceptual model, once designed, can be tested on users or other task-domain experts before further design work is done. Such testing may expose conceptual "holes" that can be corrected by revising the model, or it may expose new requirements, thereby necessitating a return to the user needs (and prioritization) stage. Even after developers decide that the conceptual model is stable enough to proceed, later design and evaluation work may expose problems in the conceptual model, requiring revisions. Indeed, gaps, problems, and unnecessary concepts in a conceptual model may be discovered in pre-release testing or even after release.

If testing exposes problems in the conceptual model, developers should go back and change it. They should resist the temptation to treat the conceptual model as "dead" after an initial UI has been designed from it. If developers don't keep the conceptual model current as they "improve" the UI design, they will regret it later, when they have no single coherent high-level description on which to base the user interface, the user documentation, training, or later system enhancements.

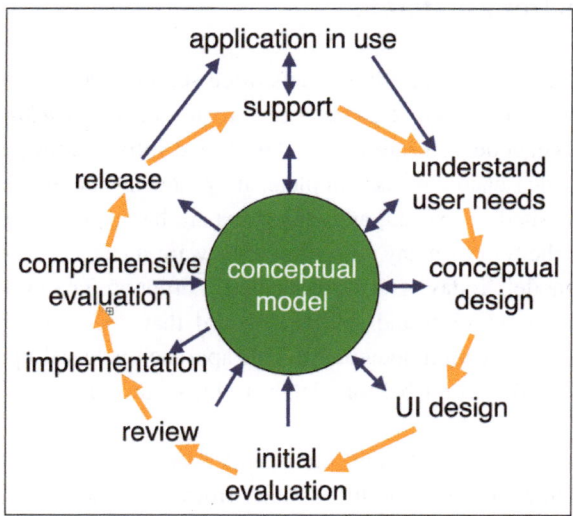

Fig. 9.3 Conceptual design's place in a user/task-centered design process

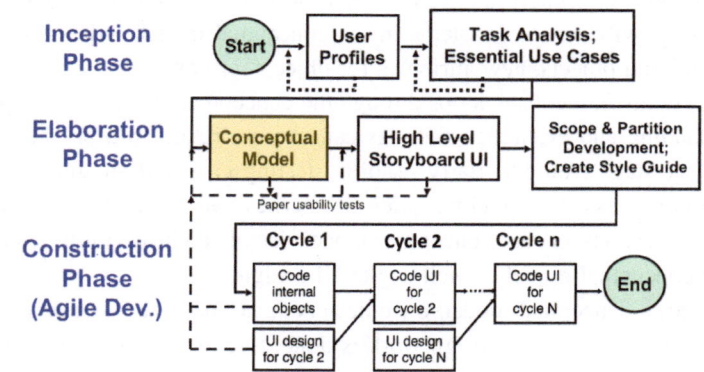

Fig. 9.4 How Conceptual Models fit into Agile development. Adapted from Jon Meads

Of course, changing the conceptual model can be painful: it affects the user interface, the documentation (in all its languages), and possibly even the implementation. The entire team is affected. But the conceptual model is the single most important part of your design. Therefore, it pays to make it as simple and task-based as possible, then do whatever you need to do to reconcile the rest of the design with it. Otherwise, the application's users will have little chance of understanding the user interface, because it will be based on a muddled conceptual model.

9.5.1 Including Conceptual Models in Agile Development

An increasing number of software development organizations use Agile methods or methods similar to Agile. Agile development involves, among other things, rapid development, and testing of successively more functional versions of the intended application (Amber 2005). Testing in Agile development makes use of team members who are task-domain experts, i.e., user representatives.[1] They provide feedback on the design during weekly or bi-weekly team "scrum" meetings. Their role is to ensure that all critical requirements are met and to keep the design focused on the tasks (Beck and Andres 2004).

Some might argue that designing conceptual models does not fit into Agile development, due to another characteristic of Agile development: its avoidance of large up-front design efforts and specification documents. Agile teams tend to prefer to specify a design in small pieces, on cards that each specify a single function or control. The idea is that the requirements and the overall UI design cannot be successfully predicted and specified, but rather will emerge as development proceeds.

The argument that conceptual models don't fit into Agile development can be countered in two ways. First, Agile and related methods were initially developed by software engineering experts, not interaction design experts, so early writings on Agile had blind spots concerning interaction design, usability, user-centered design, and how they fit into the methods. According to Bankston (2001):

> XP tends to promote a very tight focus; don't worry about what's coming, just code the card you're holding. ... Unfortunately, ... this approach can lead to disjointed, awkward or unnecessarily complicated interfaces designed around back-end functionality rather than the user's end goals. Designing an efficient and elegant user interface requires some conception of what steps comprise a given task, and how tasks interrelate to create an application's flow.

In other words, design-as-you-go methods might work for developing device drivers, operating system kernels, compilers, or other software that has little or no user interface, but design-as-you-go does not work for applications with significant user interfaces, such as airline reservation websites, air-traffic control systems, or even flight-simulation games.

In recent years, user experience experts and Agile/XP experts have attempted to "marry" their methods. One result is that they acknowledge the need for some up-front design. For example:

- Scott Ambler, an Agile/XP guru, says that the overall architecture of an application should be modeled in a "phase 0" before the normal code-test-revise cycles start (Ambler 2005). That "phase 0" model would include a conceptual model.

---

[1] Sometimes, these are actual users and sometimes they are proxies for users, e.g., developers familiar with the target task-domain from having formerly done that job.

- Larry Constantine, a UI consultant, says "some minimum up-front design is needed for the UI to be well-organized and to present users with a consistent and comprehensible interface." He and his colleague Carolyn Lockwood developed a streamlined "usage-centered" design methodology that they find fits well into Agile/XP development [2002].
- Jon Meads, a UI consultant, finds that user/task-centered design fits with Agile methods if the design (including iterations of it) occurs mainly in the Inception and Elaboration phases of a project, while the Agile code-test-revise cycles occur in the Construction phase interleaved with UI design cycles that anticipate the next coding cycle (see Fig. 9.3). One way to regard this is that iterative refinement of the conceptual design is no different than iterative refinement of the implementation code. Both are necessary.

Second, as described above and illustrated in Figs. 9.2 and 9.3, it is simply false that conceptual design only fits into a linear "waterfall" development process. User/task-centered design, of which conceptual design is a part, shares Agile's rejection of the "waterfall" model. It recognizes that customer requirements are understood better over time and change even after they are understood. It acknowledges that designs evolve but advises against beginning implementation or even detailed user-interface design until you have at least a preliminary version of a task-focused conceptual model.

On the other hand, the conceptual model, once it exists, should not be considered carved in stone. Quite the opposite, it should be tested on users and revised and improved *before* developers invest time, money, and egos on implementing code. After implementation starts, the testing and revision continue, and the design continues to evolve.

9.5.2   Testing Conceptual Models

When designing a software application, regardless of whether the development team is following an Agile process or a more conventional one, an important rule to follow is:

Test early and often.

Designers should accelerate the discovery of problems in the conceptual model by actively seeking feedback from users or other task-domain experts as early in the process as possible, before the user interface is completely designed or even begun. Focus groups can consider the concepts of the conceptual model—objects, attributes, operations, and relationships, with their terminology—and the task-flow that it implies more or less directly. Low fidelity, quick prototypes can be used to test important parts of, and issues in, the conceptual model.

Thus, lightweight usability testing can evaluate the conceptual model as well as the user interface design. In general, the earlier developers test and evaluate their conceptual model, the less likely it is that conceptual design problems will surface late in development or after release.

9.6   Getting from CM to UI

This book is mainly about conceptual modeling, but designers often ask, "How do we get from the conceptual model to the user interface?" Entire books have been written about that [see Rosenberg 2020; Jackson 2021], but for the convenience of readers, we will devote a few paragraphs to that topic here.

Once a team has developed and agreed upon a conceptual model for a planned application, the conceptual model serves as an important input for designing the application's user interfaces. "User interfaces" is plural because applications often have multiple user interfaces to accommodate different usage environments (e.g., noisy vs. quiet; indoors vs. outdoors), different device screen sizes, and devices without screens such as those for sight-impaired people or that function via voice input and output. Also, the user interface of an application includes what is presented not only via the device, but also in user documentation and training.

An application exposes—and hopefully teaches—its conceptual model to users through its user interfaces. Therefore, the UI design must represent and project the conceptual model into users' minds: it must present the objects, attributes, operations, and relationships of the conceptual model. Don Norman (2013) advocates designing user interfaces so as to narrow the "gulf of execution", which is the gulf (or gap) between users' goals and their knowledge of the operations an application provides for achieving those goals. Making the UI reflect the conceptual model is crucial to narrowing that gulf, because it allows users who have internalized the conceptual model to more easily envisage the steps needed to achieve their goal.

For example, imagine an online application for making, rescheduling, and canceling medical appointments. The conceptual model for such an application would include an **Appointment** object, and the operations on Appointments would include *make, reschedule,* and *cancel,* among others. If you have made an appointment and want to cancel it, a user interface that lists your current appointments and provides obvious ways—e.g., buttons or menu items—to reschedule or cancel them would be easier to use than, say, a user interface that lists your doctors and requires you to find the one you have an appointment with, so you can email that doctor's office to request that the appointment be canceled. The first user interface matches the conceptual model and thereby minimizes the "gulf of execution"; the second one does not.

9.6.1   Object-Based Versus Task-Based UIs

Although conceptual models are defined in terms of objects having attributes and operations, that does not necessarily mean that the best user interface will focus on the model's objects. An application's user interface can expose its conceptual model using either of two quite different approaches to interaction design: *object*-based or *task*-based. Or, as we shall see, applications can use a combination of the two approaches.

Which approach to use, or how to combine them, depends on whether people will use the application often and therefore are likely to develop a mental model approximating the designers' conceptual model, or will use the application relatively infrequently and therefore won't have a chance to develop an accurate—or any—mental model.

Object-Based Interaction: Best for Frequent Users
In many user interfaces, the objects, attributes, and operations of the application's conceptual model are exposed directly. For example, major objects of an application could be represented by titles on the menu bar (or the first layer in a phone tree) or by content objects in the middle of the page. Operations could be represented by the menu

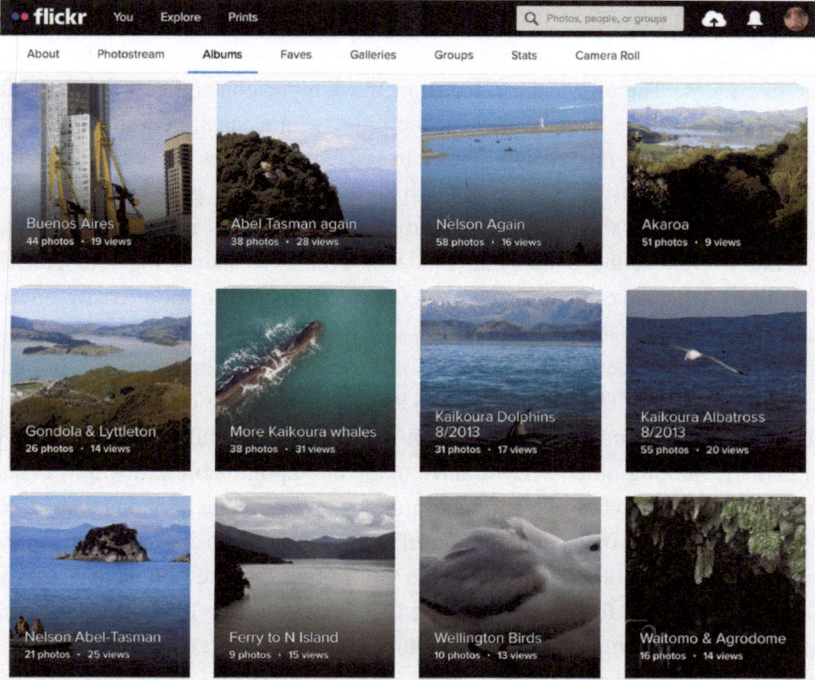

Fig. 9.5 Desktop photo management web application with object-based user interface. *Source* Flickr.com. Used with permission

Fig. 9.6 Typical Bank ATM
has task-based user interface

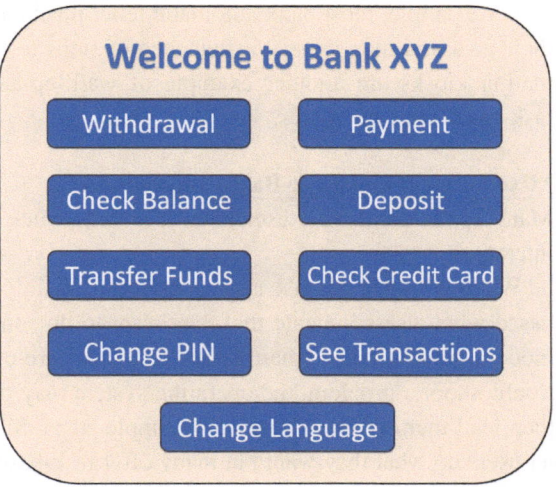

items (or the second layer in a phone tree) (see Fig. 9.5). Attributes usually appear as options or settings on objects and operations. Object-based user interfaces are often called *direct manipulation* user interfaces because they give users the illusion of being able to manipulate the objects of the conceptual model "directly".

Object-based user interfaces constantly present the application's conceptual model, helping users to internalize the model: it is immediately available by browsing through menu titles and menu items or by viewing and manipulating content objects. Even documentation for such applications is often organized around objects and operations, further reinforcing the conceptual model.

Object-based user interfaces allow users to work with whatever objects they want to, in whatever order they want to. However, that assumes that users will quickly develop a mental model of the application that at least somewhat matches the designers' conceptual model, allowing them to navigate the application freely. Therefore, object-based user interfaces are best suited for applications that users will use repeatedly and at least somewhat frequently.

Task-Based Interaction: Best for Infrequent Users

An alternative way of organizing conceptual models and applications focuses on the *tasks* (i.e., operations) that the application supports, rather than the conceptual *objects*. Some such application designs define the steps for achieving each task and provide mechanisms ("wizards") that guide users through tasks to completion. Task-based user interfaces are best for walk-up-and-use applications, where users have little or no training and don't use the application often enough to develop a mental model of how it works. Users just choose a task that matches their goal and then follow a sequence of explicitly presented steps.

For example, most bank automatic teller machines (ATMs) prompt customers to select their goal and then present a sequence of steps to the selected goal (see Fig. 6.2). Information kiosks are another example of walk-up-and-use applications that usually have task-based user interfaces.

Mixed Object and Task-Based Interaction

Many applications have user interfaces that switch between object-based and task-based interaction.

When an application's user interface is mainly object-based, it may switch to task based when users indicate that they want to do something rare or unusual. For example, modern computer and smartphone "desktops" are object based, but when a user wants to troubleshoot a problem, the operating system may prompt the user for a specific goal and then lead them through a series of simple steps. Similarly, spreadsheet programs usually let users do what they want but many offer task-based step-by-step user interfaces for rare tasks like importing data or creating elaborate graphs and charts.

Conversely, some applications start out task-based, then switch to object-based in tasks in which customers can be assumed to know what objects to operate on. For example, after a bank customer selects their desired task-goal, most ATMs list the customer's **Accounts** and ask which account the customer wants to work with. Similarly, with airline reservation websites, customers first choose a task: new reservation, change reservation, check-in for flight, manage their account, etc., but soon are working with objects: reservations, destinations, flights, seats, etc.

Benefits of Designing with Conceptual Models

10

Software development managers often balk at including a conceptual design phase in development. "How long will that take? Won't it slow us down? I don't like the idea of spending two or three weeks developing a conceptual model. Let's just start designing the screens so our implementors can get coding ASAP!".

Developing a task-focused conceptual model for an application that all major stakeholders agree on does indeed take a few weeks. However, that investment usually pays off handsomely by clarifying, focusing, and speeding up later development steps. In other words, developing a conceptual model is not simply an additional cost for a project; it produces outputs that are useful or even necessary for later steps, and that therefore can *save* development time and cost.

In this chapter, we list the downstream benefits of developing a conceptual model early in the design process.

10.1 Produces a Vocabulary

Once the development team assigns names to the conceptual objects, operations, and attributes that the application will expose to users, they have a *vocabulary* of *terms* (see Chap. 4) to be used in the application and its documentation. And agreed-upon vocabulary also provides clarity in development team discussions (see Chap. 9). As the user interface is developed, the software coded, and the documentation written, the vocabulary can be consulted to ensure that terms are used consistently throughout.

The entire team develops the vocabulary, but it is best managed and enforced by a single person, such as the product manager, the product architect, or the team's lead

© The Author(s), under exclusive license to Springer Nature Switzerland AG 2024
J. Johnson and A. Henderson, *Conceptual Models*, Synthesis Lectures
on Human-Centered Informatics, https://doi.org/10.1007/978-3-031-50852-3_10

technical writer. Whoever gets the vocabulary-manager job should keep the vocabulary up to date as the conceptual model and its concepts and terms change. The vocabulary-manager should also constantly be on the lookout for inconsistencies in what things are called in the application and its user documentation. For example:

> Hey, Anoop, it's Sergei. Got a minute? On your pages in our customer-service website, you use the term "bug report". But our agreed-upon term is "service request", remember? That's what's in the vocabulary. Where's the vocabulary? At the project's wiki. Can you please change "bug report" to "service request" on all your pages? We're running usability tests on Thursday, so I'm hoping you can make these changes by Wednesday. You will? Great, thanks.

Applications developed without a vocabulary often exhibit one or both of two common user interface "bloopers" (Johnson 2007):

1. *non-task-relevant terms*: terms for concepts not in the conceptual model, i.e., that shouldn't be exposed to users (see the Sidebar 4, below),
2. *inconsistent terminology*: which takes two forms: (a) multiple terms for a single concept (see Fig. 10.1), and (b) the same term for two or more distinct concepts, i.e., overloaded terms (see Fig. 10.2).

To ensure that all terms that users encounter in using an application are relevant to their tasks, only terms that correspond to concepts in the conceptual model should be used in an application and its documentation. Terminology that is not relevant to users' tasks should be kept entirely out of software applications and their documentation.

Furthermore, the terminology for concepts should be extremely consistent; otherwise, users—especially new ones—will be confused and will take longer to become proficient in using the application. To ensure consistency as explained in Chap. 7, the design rule that application designers should follow is:

> Same name, same thing; different name, different thing (Jarrett 2008).

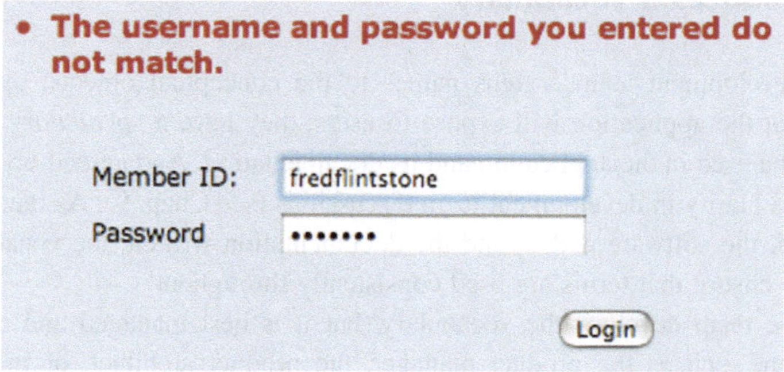

- **The username and password you entered do not match.**

Member ID: fredflintstone

Password   •••••••

Login

Fig. 10.1 The first field is labeled "Member ID", but the error message calls it "username"

Fig. 10.2 "Personal
Information" is both the name
of this whole page and of one
of its subordinate pages

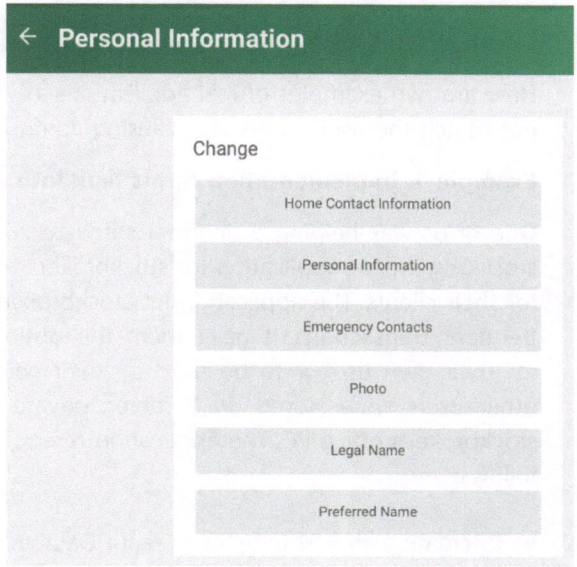

To keep foreign concepts and terms out of the user interface, the vocabulary-manager
should check for concepts and terms in the interface, software, or documentation that
aren't in the conceptual model or the vocabulary, and resist them. For example:

> Hey Sue, this screen refers to a "hyper-connector", which isn't in our conceptual model or
> vocabulary. Is it just the wrong name for something we already have in our conceptual
> model, or is it something new? If it's something new, can we get rid of it? If not, we'll
> have to take it up with the team.

In these ways, designing a conceptual model for an application facilitates a development
team's creation and use of a product vocabulary, which in turn results in an application
that is easier for people to learn and understand.

Sidebar 4: Terminology Problems in Applications

Here are two examples of real applications in which the system vocabulary did not match the user vocabulary, causing confusion.

Example 1. Implementation terms leak into user interface

One of us was helping a financial software company evaluate and improve a stock-investment application for stockbrokers to use when placing investments for their clients. The application let stockbrokers create and save templates for frequent transactions. It gave them the option of creating templates purely for their own use, or to be used by their co-workers. Templates shared with other users were stored on a server; private templates were stored on the stockbroker's office PC. The application referred to the two storage options as follows:

- To create a shared template: Create Database Template
- To create a private template: Create Local Template

The developers used a "database" for shared templates because shared templates were stored in a database. The developers used "local" for private templates on the users' own PCs because that's what "local" meant to them. The names for these two commands were changed to be more in accord with the conceptual model and more sensible to stockbrokers:

- To create a shared template: Create Shared Template
- To create a private template: Create Private Template

One could argue that this stockbroker application had one user-type that *would* need to know where transaction templates were stored: system administrators. However, one could also argue that system administrators would use an entirely different application than the one that stockbrokers used.

Example 2. Terms mean different things to users but not to the application.

Blandford and her colleagues reported a case study in which an emergency dispatch (911 in the U.S.) system's terminology did not match the terminology of the dispatchers who used it (Blandford et al. 2002).

Medical dispatchers receive reports of medical emergencies and must quickly dispatch emergency personnel and resources to handle those emergencies, often within government-mandated deadlines. Emergency situations are called "incidents". Each report of an emergency received is a "call".[1] It is common for several people to call to report the same emergency, especially since most people carry mobile phones. To get the right resources and personnel to each emergency in a timely fashion, dispatchers must know which calls are reporting which incident. They need a dispatch system that helps match calls to incidents.

The problem in a certain city was that dispatchers had to use a dispatch system that did not distinguish calls from incidents: it treated every call as a separate incident. Dispatchers had to resort to creating and sharing paper notes to try to sort out which calls were about which incident.

This case involved both types of term ambiguity. Dispatchers and the dispatch system had *different terms for the same concept*: dispatchers' term for an emergency was "incident", but the system sometimes referred to that as a "call". On the other hand, the dispatch system itself had *two terms for the same thing*: "call" and "incident" meant the same thing: an emergency.

10.2   Facilitates Creation of Conceptual Scenarios

Once the objects, attributes, and operations for an application are determined, the designers can write scenarios depicting people using the application to perform specific tasks, using only concepts and terms from the task-domain, the conceptual model, and the corresponding product vocabulary. Such scenarios are sometimes called *conceptual scenarios* or *use-cases*.

For example, for a bank account application, after a conceptual model has been developed and a product vocabulary has been agreed upon, it should be possible to write scenarios such as this:

John checks his account balance. He then deposits a check into his checking account and transfers funds from there into his savings account.

Note that this scenario refers to objects and actions from the task-domain and conceptual model only. It does not refer to details of any user interface. The scenario does not indicate whether John is interacting with a GUI on a personal computer or a voice-operated Artificial Intelligence over a telephone. However, it does specify functionality that the

---

[1] In some cities, calls are called "tickets".

application must have, as well as placing some constraints on the design. Conceptual-level scenarios are discussed in more detail in Chaps. 4 and 5, with more examples.

Conceptual scenarios are useful in checking the soundness of the design, e.g., in product functional reviews. Secondly, they can be presented directly to users in product documentation and training. Thirdly, because conceptual task scenarios describe tasks and goals without revealing the keystroke-level user actions required to achieve those goals, they can be used as task descriptions in usability tests. For example: "Please check your account balance." Finally, they provide the basis for later, more detailed scenarios written in the terminology of the eventual user interface design.

10.3   Facilitates Creation of User Documentation, Training, and Support

As mentioned above, vocabulary and task scenarios that come out of the conceptual model are important inputs to the writing of user documentation and training.

However, more importantly, a conceptual model provides those who write user documentation, training, and support materials with a high-level, conceptual view of the application. It gives them a "big picture" view of the application, which helps them write instructions for using it that make sense to users.

Furthermore, the vocabulary of the conceptual model provides a clear point of contact for people who localize the application to specific contexts. The most common such localization is translating the application into other languages. Less common, but at least as important, is creating specialized versions of the application for targeted situations. For example, corporations may have internal vocabularies that when applied to an application cause it to use the local lingo (e.g., in a law firm, folders may be called "redwelds"; in a shipping company, crates may be called "loads"). Having a single point of contact can greatly decrease the dependence of specialization activities on tracking details of the user interface design.

In addition, an application's conceptual model can be presented directly to users in its documentation, training, and support materials, thereby giving users a jump-start on forming an effective mental model.

10.4   Focuses User Interface Design: Gives Designers a Clear Target

The user interface design presents the abstract concepts of the underlying conceptual model as concrete images, controls, user actions, and task-flows. It does that whether it is based on a carefully designed conceptual model or not. In other words, the user interface is a projection of *some* conceptual model, whether that conceptual model was intended or not and whether it is coherent or not. Therefore, if designers want their application to have a coherent user interface, they should explicitly design a conceptual model for it.

An explicitly created conceptual model for an application can be regarded as a *declaration* of concepts that the application will (and will not) expose to users. It can be treated like a contract: the application's user interface can present a concept (object, operation, attribute, relationship) to users if and only if the concept is in the conceptual model.

A conceptual model gives designers a clear target for what the user interface must deliver to users. The presentation and keystroke-level interaction of the objects, operations, attributes, and relationships must be designed. The conceptual model then offers the basis for tests of how well the user interface works, that is, whether users can manipulate the objects through their UI representations as the designers intended.

The conceptual model constrains the user interface: concepts that are not in the conceptual model should not be presented by the user interface. This focuses the interface design effort, not only keeping non-task-relevant, potentially confusing concepts out of the interface, but also saving designers time and effort.

Once a user interface is designed, the conceptual task scenarios (see above) can be fleshed out with user-interface details, even down to the level of keystrokes. For example:

John double-clicks on the icon for his account to open it. A separate window opens showing the current balance. He clicks in the blank entry field below the last recorded entry and types the name and amount of a check he recently received. ...

10.5   Jump-Starts and Focuses the Implementation

Readers who are programmers will have noticed the similarity between the objects/operations analysis described here and the object-oriented analysis that is a common early step in software engineering. One big difference is that objects/operations analysis is restricted to concepts that are exposed to users while object-oriented analysis includes all concepts that are part of the implementation.

Nonetheless, having performed an objects/operations analysis provides a development team with a subset of the objects, object operations, and object attributes that must be implemented. Therefore, after the conceptual model has stabilized and agreed upon, software developers can begin implementing the internal structure of the required objects, operations, and attributes without waiting for designers to design the detailed user interface. In accordance with the Model-View-Controller architecture (Reenskaug 1979) the developers need not wait for the user interface design, i.e., the *views* and *controllers*. They can start developing the internal *models*, and the views and controllers can come later.

Using our previous example of a personal banking application: once the development team knows what types of accounts the application offers and what the operations and attributes of each type of account will be, programmers can begin coding those objects, without knowing yet how accounts will be displayed, edited, or manipulated.

Thus, contrary to the fears of many development managers, developing a conceptual model does not delay the start of coding. It actually may allow it to begin earlier. Having a conceptual model also helps coding proceed faster, because when programmers have a clear understanding of what they are coding, they can code it faster.

10.6   Supports Further Task Analysis

Although we regard task analysis as an *input* to conceptual modeling, the reverse can also be true. Early in application creation, exploratory versions of conceptual models can help guide user research and task analysis. As described in Chap. 9, the process is iterative, so objects, attributes, operations, and relationships in preliminary conceptual models can help guide user research and analysis of the user research data.

Warning: If the design team falls too much in love with its conceptual models, that framework can blind them to other understandings of what people are doing.

10.7   Supports Controlled Iteration of a Design

A conceptual model for an application should be treated as a *binding* document. If a concept is in the conceptual model, it is exposed in the application's user interface(s), and if a concept is *not* in the conceptual model, it *cannot* be exposed to users. Because the conceptual model communicates the application's high-level design to all members of the design/development team, any aspect of the emerging application that does not conform to the design as expressed in the conceptual model will likely be noticed and flagged by someone. This keeps the design true to the conceptual model and prevents implementation concepts from "leaking" into the user interface (see Chap. 9).

However, conceptual models, once developed, should not be treated as carved in stone. As Chap. 9 explains, the design process should be iterative. After a conceptual model has been developed, later design activities may expose the need to adjust it. Designing the user interface, prototyping, usability testing, implementing, writing user manuals, and developing training may expose the need for additional objects, attributes for objects, or operations on objects. As Chap. 7 explains, further task analysis may uncover "hidden" objects. Usability testing may indicate that the application's users want an attribute to have more possible values (e.g., heat: *low, medium,* and *high* instead of just *low* and *high*), or that a conceptual object needs another operation (e.g., **Photo**: *mark as favorite*). Therefore, the conceptual model should be considered a working document that can evolve.

On the other hand, changing the conceptual model should not be done lightly or often. It should be treated like a *contract* or a government *constitution*: amendable only when enough stakeholders understand the change and ratify it. For conceptual models, "enough stakeholders" means *all* members of the design/development team. Every team member can weigh the impact of the change on their respective responsibilities and deliverables. This helps prevent any team member or members from "hijacking" the design, adding features that benefit only a limited constituency.

By being both *binding* and *hard to change*, conceptual models support *controlled* design iteration.

10.8 Facilitates Communication Between Team Members

A conceptual model for an application, if well documented (see Chaps. 5 and 6), communicates the intended design to the entire design/development team. Each team member, whatever their role, can then use their understanding of the design to carry out their part of the project, whether it be designing or developing the user interface, implementing the functionality, or writing documentation. Understanding the conceptual model also gives all team members—even non-technically trained ones—the knowledge required to critique aspects of the design and make meaningful suggestions.

Having a conceptual model also facilitates the *management* of the development process and the team. It does so in two ways:

- By supporting a shared understanding of the design and thereby giving all team members a sense of agency.
- As stated in the previous section, by being *binding* and *hard to change*, conceptual models help management keep the design under control.

10.9 Saves Time and Money

A bonus benefit that results from the above-described benefits is that creating a conceptual model for an application can save time and money. Yes, creating a good conceptual model takes time: a week to a month depending on the complexity of the task-domain and the intended application.

However, due to all the above-described benefits, a conceptual model often saves development teams unnecessary effort. Without a conceptual model to guide them, user interface designers often struggle to design user interfaces that work well for all required delivery devices and user-types. Without a conceptual model, software developers often spend immense amounts of time doing Object-Oriented analysis, coding, debugging, and recoding to achieve the desired behavior.

With a carefully designed, agreed upon, and well-documented conceptual model, all team members, including user interface designers and programmers, can do their part of the project more quickly. This can *shorten* the rest of the development process by more than the time it took to develop the conceptual model.

10.10 Conclusion

In summary, the main benefits of designing a conceptual model for an application are:

- Produces a vocabulary,
- Facilitates creation of conceptual scenarios,
- Facilitates creation of user documentation, training, and support,
- Focuses user interface design: gives designers a clear target,
- Jump-starts and focuses the Implementation,
- Supports further task analysis,
- Facilitates iteration of the design, including the conceptual model,
- Facilitates communication between team members, and
- Saves time and money.

Glossary of Terms Used in This Book

The words used in this book can be seen to be two sorts: the terms (shown in *italics*) that are intended to be crisply understood (so-called "terms of art"), and those that are not (the rest of the language). Here, we show the terms used in this book that we consider important to define. The book as a whole is intended to provide a much more thorough understanding of these words as terms.[1]

activity	a step in a *plan*
application	an abstract interactive mechanism that enables a *person* to manipulate *objects* of a *domain* through its *operations*
attribute	an aspect-oriented *concept* of an *object*
component	a part of a *CM*
concept	a named abstraction used in the *structure component* of a *CM*
conceptual model (*CM*)	the designer's understanding of an *application*; the ideal *user's mental model*
conceptual scenario	a *task* with a *plan* for its performance described in terms of the *CMs* of the *applications* used in the *plan*
constraints	limitations on the structure of the conceptual model
design team	the team of designers creating an *application*
domain	subject matter of *people* using *applications; also a component of a* CM
functionality	the behavior of an application
implementation	a realization of an *application* or a *UI*, often software

---

[1] A key purpose of a conceptual model is to provide a vocabulary of terms for designers and users of the application. Hence, terms are part of the subject matter of the book. One can think of this book as a CM for CMs; this glossary would then be its vocabulary. However, to avoid confusion arising from conflating our descriptions with the things that we are describing (CMs), in this book we have tried to use the words in this glossary only as defined herein and never otherwise.

© The Editor(s) (if applicable) and The Author(s), under exclusive license
to Springer Nature Switzerland AG 2024
J. Johnson and A. Henderson, *Conceptual Models*, Synthesis Lectures
on Human-Centered Informatics, https://doi.org/10.1007/978-3-031-50852-3

interaction	all aspects of the engagement of a *person* with an application
mental model	synonym for user's mental model
metaphor	*a design based on an analogy with another tool or application, used to leverage users' prior knowledge*
model	a simulation of something that represents important aspects of the thing
object	a noun-oriented *concept*
operation	a verb-oriented *concept; something that can be done to an object*
parameter	named inputs to an operation
person	an actor performing a *task* in a *domain*
plan	*activities* by which a person can perform a *task*
presentation	all aspects of the engagement of a *UI* with a person using (and thinking about using) an *application*
purpose	the planned functionality *component* of a *CM*
relation	a relationship-oriented *concept*
resolved issues	the issue-oriented *component* of a *CM*—decided issues
service	an *application* implemented is the activity of people (cf. software)
structure	an objects/operations analysis *component* of a *CM*
sub-task	a task that is one step in a larger task
subteam	a portion of a *design team* responsible for an aspect of the design
subtype	a specialization-oriented *relation* between *types*
task	something in a *domain* that a *person* wants to get done
task domain	synonym for domain
type	a category grouping of *objects*
unresolved issues	the issue-oriented *component* of a *CM*—issues yet to be decided
use	an *activity* consisting of invoking an *application*
user interface (UI)	an interactive mechanism enabling a *person* to interact with an *application; consists of mechanisms for input and output*
user's mental model	a *person's* understanding of a *domain*, a *task*, and a *conceptual scenario* for performing it
value	the value of an *attribute* of an *object*
vocabulary	the terms defined by a *CM*; the *names* of the *concepts*.

Appendix: Origins, History, and Contributing Fields

Conceptual modeling as a tool/method to aid in designing software has a long and broad history.

Long, because as early as the 1950s and 60s, computing pioneers and researchers sought ways to specify software abstractly, in terms of datatypes, attributes, operations, relationships, and constraints, rather than the computing steps (e.g., code) required to implement it. To this day, abstract, high-level, machine-independent program specification via conceptual models remains a goal for some advocates of conceptual modeling. Others have the lesser goal of using conceptual modeling as a design tool to help ensure that software user interfaces are focused on the task-domain and the supported tasks, coherent, as simple as possible, understood, and agreed to by all team members, and easy for the intended users to learn and use.

Broad, because various aspects of conceptual modeling were invented and put into practice independently by researchers and practitioners in several subfields of computing, often with little or no awareness of work on conceptual modeling from other subfields or even from others in the same subfield. Those separate subfields include database theory and operation, business-process specification, systems analysis, simulation, software engineering, very high-level programming languages, information architecture, content management, user interface evaluation, semantic Web, task analysis and modeling, and design, particularly user interface design. Despite differences in motivation, approach, and goals, many people in these subfields independently concluded that enumerating and modeling the important **concepts** of a software system before (or instead of) coding it in a conventional programming language would facilitate development and enhancement and improve ease-of-learning and ease-of-use.

J. Johnson and A. Henderson, *Conceptual Models*, Synthesis Lectures on Human-Centered Informatics, https://doi.org/10.1007/978-3-031-50852-3

Database Design and Development, Software Engineering, and Simulation

Of the many computing subfields that have contributed to the growth of conceptual modeling, the three most central—and the earliest—are database design, software engineering, and simulation. As stated above, high-level, hardware-independent, domain-centered specification of software systems has been a "holy grail" of computing researchers and developers ever since people started writing software in the late 1940s. Similarly, developers of software simulations of complex real-world systems also needed a way to capture and represent important aspects of a domain to be simulated, even though simulation is somewhat of a special case because the simulation software to be conceptually modeled will itself be a model.

Later, after databases appeared in the 1960s, a desire to make them reliable, secure, and less ad-hoc motivated the development of formal database theory, database standards, relational and other types of databases, and the entity-relationship (ER) approach to modeling database structure and operation. A 1975 report by the ANSI/X3/SPARC Study Group on Data Base Management Systems used the term "conceptual schemas", foreshadowing the appearance of the term "conceptual models" (Sølvberg, in Olivé, 2007).

Research and development in software and database design jointly inspired the founding in 1979 of the Entity-Relationship (ER) conference as a biennial forum to advance ER modeling as a method for designing software applications involving databases. The ER conference soon became an annual event and broadened its scope beyond database applications to promote and report research on conceptual modeling of all software systems (see Table A.1: Milestones in Conceptual Modeling).

In the early 1980s, the International Standards Organization (ISO) convened a working group of software design experts—cryptically named TC97/SC5/WG5—to develop standards for software development. That group's preliminary report "Concepts and Terminology for the Conceptual Schema" (van Griethuysen et. al., 1982) is probably at least partly responsible for a significant increase in the popularity of conceptual modeling as an implementation-independent way to specify software applications in the late 1980s and early 1990s. ER charts were still popular for communicating conceptual models among development team members, but there was a desire for a more formal, machine-executable way to represent them. This led to a proliferation of modeling languages until several prominent software-engineering thought-leaders joined forces and developed the Unified Modeling Language (UML), which was released in 1995 by Rational Software.

Despite the existence of ER graphs and UML as ways to represent conceptual models, there was still significant disagreement among researchers about terminology, which aspects of an app should be modeled, and how conceptual models should be represented, prompting calls for creating a unified meta-model of proper conceptual models. Olivé (2007) and later da Silva (2015) attempted to enumerate, distinguish, and reconcile differing views of conceptual modeling within the ER research community. Nonetheless, divergent ideas about conceptual modeling persisted, as can be seen from the divergent approaches presented by contributors to Embley and Thalheim's (2011) *Handbook of Conceptual Modeling*, and in a panel discussion by five CM experts (Robinson et al., 2015) that was held at the same simulation conference where da Silva (2015) presented his literature survey supposedly reconciling those differences.

User Interface Design

A second line of research and development on conceptual models that appears to have arisen independently focused specifically on the design of software user interfaces (UIs).

In 1977, researchers and developers at Xerox's Palo Alto Research Center and Office Systems Division formed a working group to devise a systematic, repeatable method for designing the UI of Xerox's 8010 Star workstation and later software products. The method was initially documented only in an internal Xerox document, "A Methodology for User Interface Design" (Irby et al., 1977), but was later described in a publicly published retrospective by two Xerox PARC researchers, including one of the internal document's authors (Card & Moran, 1986, 1988). (We—the authors of the present book—were familiar with the original document because we worked at Xerox.) The retrospective includes the following:

> It is clear that users attempt to make sense – by building mental models – of the behavior of a system as they use it. If a simple model is not ... provided, users formulate their own myths about how the system works. ... [I]f the user is to understand the system, the system has to be designed with an explicit conceptual model that is easy enough for the user to learn. We call this the intended user's model, because it is the model the design intends the user to learn. Just what mental model the user actually forms is another issue, which depends on how clearly the intended user's model is designed, how well it is implemented, and how well it is documented.

The method involved analyzing the target task domain and developing a "conceptual model" specifying the object (types) of the domain, the attributes of each object, and the actions users could perform on each object. The concepts of the model—objects, attributes, and actions—were to be restricted to those that it was necessary to expose to users. Implementation concepts were to be excluded from the conceptual model. Also banned were concepts related to the specific presentation of the UI, so the conceptual model would be independent of any specific presentation. However, although the method stated that UI designers should model objects, attributes, and actions of the target application domain, it did not describe in detail how to do so.

In the 1988 first edition of his book *The Design of Everyday Things* (originally titled *The Psychology of Everyday Things*), Don Norman explained the importance of designing

UIs of an appliance or software application to embody and project a task-focused, simple—even over-simplified—conceptual model so its users develop a mental model that facilitates learning, retention, and reasoning about how to operate it. He states:

> A conceptual model is an explanation, usually highly simplified, of how something works. It doesn't have to be complete or even accurate as long as it is useful.

Later editions expanded the discussion of conceptual models but didn't provide a method for developing them and sometimes blurred the distinction between conceptual models, users' mental models, and implementation (Norman, 2013).

Also in the 1990s, software engineer Dave Collins, drawing upon his experience designing and developing software as well as his readings in human cognitive psychology, user interface design, and software engineering, wrote a book on what he called "object-oriented user interfaces". He asserted that using object-oriented design for both the user interface and the underlying software "brings coherence to the software development process", as well as producing user interfaces that make more sense to users. The book lays out a method for user interface design that includes identifying and analyzing the important objects of the task domain and their relationships and performing a task analysis. The results of the object and task analyses are used to design a conceptual model for the application, which is the model the designers want users to adopt as their mental model. The conceptual model then guides the design of the concrete user interface (presentation and interaction), so the user interface "teaches" users the desired conceptual model. In Collins' method, the conceptual model also guides the object-oriented implementation of the software—specifically the objects that users manipulate.

During that same period, we—the present authors—as practitioners of UI/UX design at various companies after Xerox, applied what we had learned at Xerox and from Norman about the value of conceptual models as a UI design tool. Because the Xerox document and Norman's book provided little advice on how to **do** conceptual modeling, and because we were unaware of the ER conferences and the prior research on conceptual modeling by software and database developers, we developed our own methods for identifying the concepts of an application domain and constructing a model based on them. In the early 2000s, we jointly decided that more practicing UI/UX designers needed to know the value of conceptual models and how to construct them. Unaware of Collins' 1995 book on that topic, we wrote an article for the ACM SIGCHI magazine *Interactions* (Johnson and Henderson, 2002). Based on positive feedback about that article and on what we learned from applying conceptual modeling to later UI design projects, we expanded that article into a book (Johnson and Henderson, 2011).

In the 2000s and 2010s, Prof. Ann Blandford and her colleagues published a series of studies and investigations about evaluating the fit between the conceptual models of interactive systems and the mental models of those systems' users (Blandford et al, 2008;

Blandford, 2013). These papers describe methods to elicit peoples' mental models, determine the conceptual models of existing interactive systems, check how well those models match (or don't match) each other, and thereby discover potential design improvements to remedy the conceptual mismatches.

Recently, Daniel Jackson, a software engineer and computer science educator with an interest in UI design, wrote a book *The Essence of Software: Why Concepts Matter for Good Design* (Jackson, 2021), urging UX/UI and software designers to identify and understand an application's key **concepts** and the relationships between them before sketching or prototyping the application's screens and controls. The book is based on the author's readings on UI design and conceptual modeling, as well as on his observations of people—including himself—using and often struggling with software. It presents many examples of poor conceptual design—where an application's concepts don't match users' mental models so it sometimes behaves unexpectedly—and explains how the application might be simplified or redesigned to convey its concepts more clearly to users. Jackson then presents a method for designing applications based on concepts. The method is similar to the standard object-oriented approach to conceptual modeling but differs in that the defining concepts of an application need not be objects; they can also be actions (operations in this book's lexicon) or relationships.

User Interface Engineering and User Interface Management Systems

A third computer science subfield that influenced thinking about conceptual models started in 1979 with Mikkjel H. Reenskaug's invention of the Model-View-Controller (MVC) software design pattern (Reenskaug, 1979). According to Reenskaug:

> The top level goal was to support the user's mental model of the relevant information space and to enable the user to inspect and edit this information.

Software developers, including the developers of the SmallTalk object-oriented programming language (Krasner and Pope, 1988), quickly recognized the value of modeling the abstract logic behind user interfaces. For example, a UI for ordering a pizza might be modeled abstractly as an enumerated choice of pizza sizes, a set of ON/OFF switches for optional toppings, another ON/OFF switch for choosing pickup or delivery, and ways to input delivery addresses, phone numbers, and payment card numbers. With the semantics of the UI abstracted in this way, the actual presentation (i.e., view) and interaction methods (i.e., controllers) of the UI could depend on delivery modality (e.g., graphical, textual, voice), device characteristics (e.g., screen size), user accessibility requirements (e.g., blind, slow-keying, unsteady hands), and other factors (e.g., task).

Starting in the late 1980s, researchers and developers of advanced User Interface Management Systems (UIMSs) such as CMU's Garnet and Amulet (Myers, 1988), Georgia

Tech's UIDE (Foley et al, 1988), IBM's ITS (Gould et al, 1991), and HPLabs' ACE (Johnson et al., 1993) used MVC architecture to separate application front ends into layers: semantics, presentation, and control. However, the models in the MVC architecture and in these UI management systems are models of the **user interface**, not of the application semantics.

Later, USC's HUMANOID (Szekely et al, 1993) application development environment extended the use of MVC to model application semantics. Previous UI builders and UIMSs focused on specifying the application **controls** (aka settings), but HUMANOID also allowed designers to abstractly specify an application's main content objects—e.g., **organizations** in an org-chart application—and the effects that user operations have on those content objects. However, HUMANOID was limited in the types of content objects that could be specified and did not support designing UIs in which users could add new types of content objects.

Another approach to facilitating user-interface development is to create domain-specific user interface development tools and environments. An example of this was Trillium, a design environment for developing user interfaces for photocopiers (Henderson, 1986).

Task Analysis and Task Models

Developing an application requires understanding domains in which it is used, which sorts of people will be using it, what tasks they are performing, and how the application will fit into the plans for performing those tasks. Since long before the advent of digital computers, human behavior has been an object of interest in the social sciences, design, computer sciences, and business. Broadly, there have been three intertwined accounts of human activity:

- *Activity Theory* emerged from psychologists' efforts to understand how people's consciousness interacts with the context in which their activity takes place, and in particular, people's interaction with objects (Vygotsky, 1929; Leont'ev 1974). Activity theory was first applied to computer system development in Scandinavia and then more broadly [see Carroll (1991), Nardi (1996)].
- *Situated Action* grew out of anthropology. In the early 1980s at Xerox PARC, studies of people working in offices and delivering services found that plans are central to interaction between human activity and applications [see Suchman (1987, 2007)].
- *Distributed Cognition* arose from cognitive science. It postulates that intelligence is manifested not at an individual level, but rather in communication between entities (teams of people and computer applications) [see Yates (1989), Hutchins (1994)].

These approaches to task analysis share the following viewpoints:

- The subject of analysis should be *naturally occurring* activity. Consequently, all three accounts involve deep engagement between task analyzers and those performing tasks. Analysis of this activity is the essential input data for developing conceptual models of applications.
- The activity under analysis is richer than any analysis of it [see Geertz (1973)]. As part of using applications, users bridge the gap between the conceptual model and the world [see Henderson (2008)]).
- The activity of performing tasks often changes people's understanding of tasks, their plans for accomplishing tasks, and the domain.

Activity theory, situated action, and distributed cognition were not originally concerned with the design of computer software applications. However, they have been applied to that purpose by researchers and software design experts attempting to develop better methods of analyzing and representing the tasks for which people use computer applications. They have become part of the foundation of task analysis. Examples of published works on task analysis include the following:

- Beyer and Holzblatt (1996) developed and published *Contextual Design*, a method of conducting user analysis and task analysis and representing the results to inform application design (see Fig. 9.2), and updated it twenty years later (Beyer and Holzblatt, 2016].
- In 1991, Joan Greenbaum and Morten Kyng presented *Design at Work*, an edited collection of papers resulting from a workshop between researchers of the Scandinavian System Development movement (Greenbaum and Kyng (eds.), 1991].
- In the early 2000s, Harvard Business School professor Clayton Christensen and Deloitte Research Director Michael Raynor urged companies to identify "jobs" that customers will buy products to help them with.
- In 2008, software strategy consultant Indi Young published *Mental Models: Aligning Design Strategy with Human Behavior* (Young, 2008) providing detailed methods of interviewing users, extracting domain objects and operations, and proposing applications.
- A year later, Tim Brown, CEO of design consultancy IDEO, published *Change by Design* arguing that by following best practices, design can be done in a relatively linear fashion, reducing the amount of iteration required (Brown, 2009).
- Building on the work of Christensen and Raynor, Jim Kalbach of Harvard Business School published *The Jobs to be Done Playbook*, focusing once again on the importance of tasks in design (Kalbach, 2020).
- Most recently, Judy Bowen and her colleagues conducted a comprehensive review of the literature on task models (Bowen et al., 2021), which are representations of the defining aspects of a task.

Information Architecture

Interest in conceptual modeling—or variants thereof—also arose within the Web Information Architecture (IA) field, because some in that field saw a need to represent content elements and their connections in Web sites and Web applications independently of page designs. For example, Web content strategist Rachel Lovinger (2012) wrote, in an article posted on the IA blog-site *A List Apart*:

> The content model is one of the most important content strategy tools. It allows designers to represent content in a way that translates the intention, stakeholder needs, and functional requirements from the user experience design into something that can be built by developers implementing a CMS [content management system].

In Lovinger's method, a content model documents all types of content in a Web site or Web application, including the relationships between elements. For example, a content model for a pop music website could include objects such as songs, composers, artists, albums, and song-popularity lists, with appropriate links between them. Lovinger described how to build content models and recommended representing them as entity-relationship (ER) diagrams or as spreadsheets, with the level of detail determined by purpose.

Soon after Lovinger's article was posted, the IA blog-site *Boxes & Arrows* invited us to contribute an overview of Conceptual Models aimed at their audience (Johnson & Henderson, 2013).

Two years later, *A List Apart* posted Sophia Prater's article "Object-Oriented UX" (Prater, 2015) advocating her variant of conceptual modeling. As a UX designer for several very complex, high-profile websites, Prater developed a design method based on "content-out design, not canvas-in design". Her method builds on what she learned from a book by Dave Collins titled *Designing Object-Oriented User Interfaces* {2008]. It involves first identifying the real-world objects of the task-domain: "We determine the actions after first defining the objects, as opposed to the traditional actions-first process that jumps straight into flows, interactions, and features." Collins and Prater claim that using OO design for front ends matches how developers work, and so "brings coherence to the software development process". Prater explained OOUX further in an interview on the IA podcast: *The Informed Life* (Arango, 2021).

The architects of the World Wide Web have since the Web's inception worked to develop ways for Web designers to specify different aspects of a website—e.g., content, meaning, presentation style—separately. One result of those efforts is cascading stylesheets (CSS) (Lie and Bos, 1996; W3C, 2023; Pemberton, 2023), which allows website *content* to be specified separately from its *presentation*. Further efforts led to the Extensible Markup Language (XML), which allows Web designers and developers to define and combine *medium-specific* content specification languages (W3C, 2008; Pemberton, 2023), and the Resource Description Framework (RDF) (Pemberton, 2009) and

Web Ontology Language (OWL), which support adding machine-readable metadata to website elements to declare their *meaning*. Taken together, these efforts allow Web content creators to specify the content, presentation, and machine-readable meaning of websites and web applications separately (Table A.1).

Table A.1 Milestones in conceptual modeling

Year	Database & Software Eng	User Interface Design	UI Management Systems	Task Analysis and Task Models	Info Architecture & Semantic Web
1925				Psychologists studying human behavior and consciousness (Vygotsky]	
1950s	Desire arises for abstract, formal spec of software				
1960s	Databases appear				
1967				Garfinkel's ethnomethodology studies of situated action	
1970	Codd invents Relational DBs				
1973				Anthropologist Geertz introduces "thick description"	
1974				Activity theory studying human behavior (Leont'ev]	
1975	ANSI doc on DBs uses term "Conceptual schema"				
1976	Chen invents Entity-Relationship (ER) modeling of DBs				
1977		Irby et al. Xerox internal doc recommends conceptual modeling for UI design; CMs used in UI design of Xerox 8010 Star workstation			

(continued)

Table A.1 (continued)

Year	Database & Software Eng	User Interface Design	UI Management Systems	Task Analysis and Task Models	Info Architecture & Semantic Web
1979	IBM releases first commercial relational DB. First ER/CM conference		M. Reenskaug invents model-view-controller (MVC) architecture; It is immediately adopted in Smalltalk and elsewhere		
1981	Second ER/CM conference	Xerox 8010 Star released; its UI embodies a highly coherent CM based on desktop metaphor			
1982	ISO doc mentions "Conceptual schemas" and promotes CMs				
1983	Third ER/CM conference	Card, Moran, and Newell discuss CMs in *Psych of HCI* book			
1985	Fourth ER/CM conference; continues annually				
1986		Card and Moran recount their early use of CMs	Henderson describes Trillium copier-UI development environment		
1987				Suchman publishes book *Plans and Situated Action*	
1988		Norman argues for CMs in POET/ DOET book	UIMS systems use MVC to facilitate UI design (continues through present)		
1989				Bødker's article "A Human Activity Approach to User Interfaces" published	

(continued)

Table A.1 (continued)

Year	Database & Software Eng	User Interface Design	UI Management Systems	Task Analysis and Task Models	Info Architecture & Semantic Web
1991				Activity Theory meets Situated Action: *Design at Work* published	
1994				Hutchins publishes book *Cognition in the Wild* on distributed cognition	
1995	Rational develops UML	Collins promotes CMs in book on OOUI design			
1996					W3C releases CSS
2002		Johnson and Henderson *Interactions* article promotes CMs for UI design			
2003				Christensen and Raynor's *The Innovator's Solution* published	
2006					W3C releases XML 1.0
2007	Olivé's book *Conceptual Models in Information Systems* surveys work on CMs within ER/CM community and attempts to organize and reconcile it			Suchman updates *Plans and Situated Action*, renaming it: *Human–Machine Reconfigurations: Plans and Situated Action*	
2008		Blandford et al. paper on CASSM method for evaluating system utility and conceptual fit		I. Young's book *Mental Models* published	
2009				Brown's book *Change by Design* published	Pemberton's W3C tutorial on RDF posted; W3C releases OWL 2.0

(continued)

Table A.1 (continued)

Year	Database & Software Eng	User Interface Design	UI Management Systems	Task Analysis and Task Models	Info Architecture & Semantic Web
2009		Johnson and/or Henderson present CM courses at ACM CHI conference (2009–2015 and 2019)			
2011	Embley and Thalheim publish *Handbook of Conceptual Models*	Johnson and Henderson publish book *Conceptual Models*			
2012					Lovinger's "Content Modeling" article posted on IA site *A List Apart*
2013		Norman updates DOET book and adds more about CMs. Blandford paper on eliciting users' mental models published			Johnson and Henderson's article on CMs posted on Boxes & Arrows IA blog
2015	Panel on CMs at Winter Simulation Conference presents four different views of CMs da Silva's book chapter "Model-driven Engineering" surveys work on CMs within ER/CM community and attempts to organize and reconcile it				Prater's article "Object-Oriented UX", posted on IA site *A List Apart*
2020				Kalbach's *Jobs to Be Done Playbook* published	
2021		Jackson's book *The Essence of Software: Why Concepts Matter for Good Design* published		Bowen et al. literature survey of task-modeling research	Arango interviews Prater about OOUX on Informed Life IA podcast

Bibliography

Abowd, G and Dix, A. (1992). "Giving undo attention" Interact. Comput. 4, 3 (December 1992), 317–342.

Ambler, S., (2005). The Agile System Development Lifecycle. Ambysoft online article, http://www.ambysoft.com/essays/agileLifecycle.html

Arango, J. (2021) Interview of Sophia Prater on Object-Oriented UX, The Informed Life Podcast, June 6, Online article: https://theinformed.life/2021/06/06/episode-63-sophia-prater/

Bangston, A. (2001), "Usability and User Interface Design in XP." Online article: http://www.ccpace.com/Resources/documents/UsabilityinXP.pdf .

Beck, K. and Andres, C. (2004), *Extreme Programming Explained: Embrace Change, 2nd Ed.* Reading, MA: Addison Wesley.

Beyer, H. (2010), *User-Centered Agile Methods*, Morgan & Claypool.

Beyer, H., and Holtzblatt, K. (1997) *Contextual Design: Defining Customer-Centered Systems 1st ed.* Morgan Kaufmann.

Beyer, H., and Holtzblatt, K. (2016) *Contextual Design: Design for Life, 2nd ed.* Morgan Kaufmann.

Bittner, K. & Spence, I. (2003). Use Case Modeling. Addison-Wesley. p. xvi.

Blandford, A. E., Wong, B. L. W., Connell, I. & Green, T. R. G. (2002) "Multiple Viewpoints on Computer Supported Team Work: A Case Study On Ambulance Dispatch". In X. Faulkner, J. Finlay & F. D!étienne (Eds.) *People and Computers XVI: Proc. HCI'02.* 139–156. Springer.

Blandford, A. (2013) "Eliciting People's conceptual Models of activities and Systems", *International Journal of Conceptual Structures and Smart Applications*, 1(1), 1–17, January–June 2013.

Blandford, A., Green, T. R. G., Furniss, D. & Makri, S. (2008) "Evaluating system utility and conceptual fit using CASSM". *International Journal of Human–Computer Studies.* 66. 393–409. DOI https://doi.org/10.1016/j.ijhcs.2007.11.005

Blomberg, J, Darrah, C. (2015) An Anthropology of Services: Toward a Practice Approach to Designing Services. Morgan & Claypool.

Bodker, S. (1989) A human activity approach to user interfaces. Human-Computer Interaction 4:171–195.

Bowen, J., Ditmar, A., Weyers, B. (2021), Task Modelling for Interactive System Design: A Survey of Historical Trends, Gaps and Future Needs, Proc. ACM Human-Computer Interaction, June 2021, Vol. 5, No. EICS, Article 214.

Brown, T. (2009) Change by Design. New York, NY: HarperCollins.

Buxton, B. (2007) *Sketching User Experiences: Getting the Design Right and the Right Design.* Morgan Kaufmann.

Card, S. (1993), Discussant comments, at HCI Consortium, Boulder ,CO,

© The Editor(s) (if applicable) and The Author(s), under exclusive license to Springer Nature Switzerland AG 2024
J. Johnson and A. Henderson, *Conceptual Models*, Synthesis Lectures on Human-Centered Informatics, https://doi.org/10.1007/978-3-031-50852-3

Card, S. (1996) "Pioneers and Settlers: Methods Used in Successful User Interface Design", in M. Rudisill, C. Lewis, P. Polson, T. McKay (eds.), *Human-Computer Interface Design: Success Cases, Emerging Methods*, Real-World Context, Morgan Kaufmann.

Card, S., Moran, T., Newell, A., 1983. *The psychology of human-computer interaction.* Lawrence Erlbaum Associates, Hillsdale, NJ.

Card, S. & Moran, T. (1988) "User Technology: From Pointing to Pondering", Proceedings of the ACM Conference on the History of Personal Workstations, 1986, p. 183–198; also in Goldberg, A. (1988) [ed.], A History of Personal Workstations, ACM Press, DOI: https://doi.org/10.1145/61975

Carroll, J. M. (ed.) *Designing Interaction; Psychology at the Human-Computer Interface.* Cambridge, MA: MIT Press.

Christensen, C. M & Raynor, M.E. (2003), "What products will customers want to buy", Chapter 2 in *The Innovator's Solution, 1st ed.*, Harvard Business School Publishing.

Collins, Dave (1995), *Designing Object-Oriented User Interfaces*, Benjamin-Cummings, 590 pages.

ConceptualModeling.org, accessed Sept 2023.

Constantine, L. (2002), "Process Agility and Software Usability: Toward Lightweight Usage-Centered Design", *Information Age*, Aug–Sep.

Cooper, A. (2004) *The Inmates are Running the Asylum: Why High Tech Products Drive Us Crazy and How to Restore the Sanity.* Pierson/SAMS.

Dubberly, H. (1999) "Understanding Internet Search", Dubberly Design Office, http://www.dubberly.com/concept-maps/understanding-internet-search.html

Dubberly, Hugh. (2010) Creating Concept Maps, Report, Dubberly Design Office.

Embley, D.W. and Thalheim, B. (eds) (2011) *Handbook of Conceptual Modeling*, Springer.

Foley, J., Gibbs, C., Kim, W.C., and Kovacevic, S. (1988). "A Knowledge-based User Interface Management System." *Proceedings of the ACM Conference on Computer-Human Interaction*, 1988, ACM Press, 67–72.

Garfinkel, H. (1967), *Studies in Ethnomethodology.* Englewood Cliffs, NJ: Prentice-Hall.

Geertz, C. (1973), Thick Description: Towards an Interpretive Theory of Culture in The Interpretation of Cultures. Basic Books.

Gould, J.D., Boies, S.J., and Lewis, C. (1991), "Making Usable, Useful, Productivity-Enhancing Computer Applications, Communications of the ACM, 34(1), January 1991, 74–85.

Greenbaum, J, Kyng, M. (1991) *Design at Work: Cooperative Design of Computer Systems.* Hillsdale, NJ: Lawrence Erlbaum Associates.

Halasz, F., and Moran, T. (1982), "Analogy Considered Harmful", Proceedings of CHI 1982, pages 383–386.

Henderson, A. (1986) "Trillium: A Knowledge-based Design Environment for Control/Display Interfaces", in *Proceedings of the ACM SIGCHI conference on Human Factors in Computing Systems (CHI '86)*, Boston, MA, USA. 1986.

Henderson, A. (2008) *The 100% Solution: what is a user to do, and how are we helping?* in ECCE '08: Proceedings of the 15th European conference on Cognitive ergonomics: the ergonomics of cool interaction, https://doi.org/10.1145/1473018.1473021

Henderson, A., and Kyng, M. (1991), "There's no place like home: Continuing Design in Use", in J. Greenbaum, M. Kyng, (eds.) *Design at Work: Cooperative Design of Computer Systems.* Hillsdale, NJ: Lawrence Erlbaum Associates, pp 219–240.

Hutchins, E. (1994) *Cognition in the Wild.* Cambridge, MA: MIT Press.

Irby, C., Bergsteinson, L., Moran, T., Newman, W., Tesler, L. (1977), A Methodology for UI Design, Xerox internal document.

Isaacs, E., and Walendowski, A. (2001) *Designing from Both Sides of the Screen: How Designers and Engineers Can Collaborate to Build Cooperative Technology*, SAMS.

Jackson, D. (2021). *The Essence of Software: Why Concepts Matter for Good Design*, Princeton University Press, 323 pp.

Jarrett, C. (2008), *Forms That Work*, Morgan Kaufman Publishers.

Johnson, J. (1985), "Calculator Functions on Bitmapped Computers", *SIGCHI Bulletin*, July 1985, pages 23–28.

Johnson, J., (1987), "How Faithfully Should the Electronic Office Simulate the Real One", *SIGCHI Bulletin*, July 1987, pages 21–25.

Johnson, J (2020), *Designing with the Mind in Mind, 3rd ed.*, Morgan Kaufman Publishers.

Johnson, J. (2007) *GUI Bloopers 2.0: Common User Interface Don'ts and Dos*, Morgan Kaufmann Publishers.

Johnson, J., Roberts, T., Verplank, W., Smith, D.C., Irby, C., Beard, M., and Mackey, K. (1989), "The Xerox Star: A Retrospective", *IEEE Computer*, September, 1989, pages 11–29.

Johnson, J., Nardi, B., Zarmer, C, Miller, J.R. (1993), "ACE: Building Interactive Graphical Applications", Communications of the ACM, April 1993, 36(4), pages 41–55

Johnson, J. and Henderson, D.A. (2002), "Conceptual Models: Begin by Designing What to Design", *Interactions*, Jan–Feb 2002, pages 25–32

Johnson, J. & Finn, K. (2017) *Designing User Interfaces for an Aging Population* (Morgan Kaufman.

Kalbach, Jim. (2020) *Jobs to be done playbook*, Rosenfeld Media, New York, NY.

Krasner, G.E., Pope, S.T. (1988). "A description of the model–view-controller user interface paradigm in the smalltalk-80 system", *Journal of Object Oriented Programming.*, 1988;1:41.

Leont'ev, A.N. (1974) *The problem of Activity in Psychology*. Soviet Psychology 13(2):4–33.

Lie, H.W. and Bos, B. (eds.) (1996), Cascading Style Sheets, Level 1, W3C. Web: https://www.w3.org/TR/CSS1/

Lombardi, V. (2008) "Concept Design Tools", Digital Web Magazine, Sept. 30. Web: http://www.digital-web.com/articles/concept_design_tools/

Lovinger, Rachel (2012), "Content Modeling: A Master Skill", A List Apart (Information Architecture blog-site), April 24, 2012, https://alistapart.com/article/content-modelling-a-master-skill/

Moran, T.P. (1983) "Getting into a system: External-internal task mapping analysis", In *Proceedings of the ACM SIGCHI conference on Human Factors in Computing Systems (CHI '83)*, Ann Janda (Ed.). ACM, New York, NY, USA, 45–49.

Myers, B.A. (1988). Tools for Creating User Interfaces An Introduction and Survey." Technical Report CMU-CS- 88–107, 1988, CMU.

Nardi, B. A. (ed.) (1996a) *Context and Consciousness; Activity Theory and Human-Computer Interaction*. Cambridge, MA: MIT Press.

Nardi, B. A. (ed.) (1996b) "Studying Context: A Comparison of Activity Theory, Situated Action Models, and Distributed Cognition," in Nardi, B. A. (ed.) (1996a) *Context and Consciousness; Activity Theory and Human-Computer Interaction*. Cambridge, MA: MIT Press.

Newman, W. and Sproul, R. (1973). *Principles of Interactive Computer Graphics*. New York: McGraw–Hill.

Norman, D.A., and Draper, S.W. (1986). *User Centered System Design: New Perspectives on Human-Computer Interaction*, Hillsdale, New Jersey: CRC.

Norman, D.A., (1983). Design rules based on analysis of human error. *Communications of the ACM* 26 (4), 254–258.

Norman, D.A. (1988, revised 2013). *The Design of Everyday Things*. Basic Books.

Norman, D.A. (2010) *Living with Complexity*. MIT Press.

Olivé, A. (2007) *Conceptual Modeling of Information Systems*, Berlin, Springer-Verlag.

Ostrom, E. (1990). *Governing the Commons: The Evolution of Institutions for Collective Action*, Cambridge University Press.

Pemberton, S. (2009), "RDFa for HTML Authors", W3C, Version date: 2009–05–14, https://www.w3.org/MarkUp/2009/rdfa-for-html-authors.

Pemberton, S. (2023), "The One Hundred Year Web". In Proceedings of ACM Web Conference, ACM, New York, NY, USA, 8 pages. https://doi.org/10.1145/3543873.3585578

Prater, S. (2015), "Object-Oriented UX", A List Apart (Information Architecture blog-site), October 20, 2015: https://alistapart.com/article/object-oriented-ux/

Reenskaug:, T. (1979), "Models - Views – Controllers". Technical note, Xerox PARC, December 1979.

Robinson, S., Arbez, G., Birta, L.G., Tolk, A., Wagner, G. (2015) "Conceptual Modeling: Definition, Purpose and Benefits", in L. Yilmaz, W. K. V. Chan, I. Moon, T. M. K. Roeder, C. Macal, and M. D. Rossetti, (eds.), Proceedings of 2015 Winter Simulation Conference.

Rodrigues da Silva, A. (2015), "Model-driven engineering: A survey supported by the unified conceptual model", *Computer Languages, Systems & Structures* 43, 139–155.

Rosenberg, D. (2020), *UX Magic*, Interaction Design Foundation.

Staggers & Norcio (1993), "Mental Models: Concepts for human-computer interaction research", Int. J. Man-Machine Studies, 38, 587–605.

Suchman, L. (1987) *Plans and Situated Actions: the problem of human-machine communication* New York, NY: Cambridge University Press.

Suchman, L. (2007) *Human-Machine Reconfigurations: Plans and Situated Actions, 2nd Edition.* New York, NY: Cambridge University Press.

Szekely, P., Luo, P., Neches, R. (1993), "Beyond Interface Builders: Model-Based Interface Tools", CHI '93: Proceedings of the INTERCHI '93 Conference on Human Factors in Computing Systems, May 1993, pp 383–390, https://doi.org/10.1145/169059.169305

Vygotsky, L. S. (1925) "Consciousness as a problem in the psychology of behaviour", in *Collected Works: Questions of the Theory and History of Psychology.* Moscow: Pedagogika.

W3C (2008), Extensible Markup Language (XML) 1.0 (5th Ed.), https://www.w3.org/TR/xml/.

W3C (2023), CSS Snapshot 2023, https://www.w3.org/TR/CSS.

Washizaki, H. and Fukazawa, Y. (2002) "Dynamic hierarchical undo facility in a fine-grained component environment." in Proceedings of the Fortieth International Conference on Tools Pacific: Objects for internet, mobile and embedded applications (CRPIT '02). Australian Computer Society, Inc., Darlinghurst, Australia, Australia, 191–199.

Wikipedia, (undated) "Conceptual Model", accessed Sept 2023: https://en.wikipedia.org/wiki/Conceptual_model

Wikipedia, (undated) "Semantic Web", accessed Sept 2023: https://www.w3.org/TR/owl2-overview/

Yates, J. (1989) *Control through Communication: The Rise of System in American Management.* Baltimore, MD: The Johns Hopkins University Press.

Young, R.M. (1981) "The Machine Inside the Machine: Users' Models of Pocket Calculators." *International Journal of Man-Machine Studies* (1), pages 51–85.

Young, I. (2008) *Mental Models: Aligning Design Strategy with Human Behavior*, Rosenfeld Media, New York, NY., 291 pp.